Science for the Seventies
Book 2

TEACHERS' GUIDE

Publications for the Scottish Integrated Science Syllabus

Published by Heinemann Educational Books Ltd

Science Worksheets

Prepared by the Working Party on Secondary School Science

YEAR ONE

YEAR TWO

Available either in
(a) single-pupil *collated sets* or
(b) *multiple sets* for one hundred pupils

Science for the Seventies

Teachers' Guides

For use with *Science Worksheets* and *Science for the Seventies* pupils' texts

BOOK ONE
BOOK TWO
BOOK THREE

Pupils' Textbooks
BOOK ONE
BOOK TWO
BOOK THREE

Published by Her Majesty's Stationery Office, for the Consultative Committee on the Curriculum, Scottish Education Department

Science for General Education; No. 7 of the series Curriculum Papers

Science for the Seventies
Book 2

TEACHERS' GUIDE

Designed for use with the **Science Worksheets: Year Two,** *prepared by the Working Party on Secondary School Science in Scotland and with* **Science for the Seventies: Book 2** Pupils' Textbook

A. J. Mee
Patricia Boyd
Biology Teacher, Boroughmuir Secondary School, Edinburgh

David Ritchie
Deputy Headmaster and Principal Physics Teacher, Balwearie School, Kirkcaldy

Heinemann Educational Books
London and Edinburgh

Heinemann Educational Books Ltd

London Edinburgh Melbourne Auckland Toronto
Singapore Hong Kong Kuala Lumpur
Ibadan Nairobi Johannesburg
New Delhi

ISBN 0 435 57573 2

© A. J. Mee, P. Boyd, and D. Ritchie, 1972

First published 1972
Reprinted 1972

Published by Heinemann Educational Books Ltd
48 Charles Street, London W1X 8AH

Printed in Great Britain by
Morrison and Gibb Ltd, Edinburgh and London

The Functions of this Guide

This Teachers' Guide has two functions:

1. As a guide to Year Two of the *Science Worksheets*.
2. As a guide to Book Two of *Science for the Seventies*.

Cross-reference tables, linking this Guide and the Pupils' Textbook with the *Science Worksheets*, appear in Appendix 2 at the end of this book.

Acknowledgements

The Report of the Working Party, *Science for General Education*, *Curriculum Papers* 7, is published by Her Majesty's Stationery Office, Edinburgh. It provides the essential background to the scheme for anyone contemplating its adoption. The general comments in the Introduction to the present book draw freely on *Curriculum Papers—7* and the authors are grateful to H.M.S.O. for permission to quote from it.

Members of the Working Party on Secondary School Science in Scotland

A. J. MEE, H.M.I., Scottish Education Department (*Chairman*)

J. I. ALLAN, Dundee College of Education

J. R. BARR, Science Adviser, Edinburgh Education Authority

S. CONNELL, Assistant Science Teacher and House Mistress, Kinning Park Secondary School, Glasgow

M. FLEMING, Assistant Science Teacher, Johnstone High School

J. GEMMELL, Science Adviser, Glasgow Education Authority

P. HAAS, Assistant Science Teacher, St. Patrick's High School, Coatbridge

J. G. M. HALLIDAY, Formerly Principal Teacher of Science, Gourock High School, now Assistant Director of Education, Renfrewshire

M. J. HAY, H.M.I., Scottish Education Department

A. McINTYRE, Formerly Principal Teacher of Science, now Headmaster, Niddrie Marischal Secondary School, Edinburgh

J. MILLER, Vice-Principal, Jordanhill College of Education, Glasgow

J. MUIR, Principal Teacher of Science and Depute Headmaster, Riverside Secondary School, Stirling

D. RITCHIE, Principal Teacher of Physics and Depute Headmaster, Balwearie Secondary School, Kirkcaldy

W. R. RITCHIE, H.M.I., Scottish Education Department

S. T. S. SKILLEN, H.M.I., Scottish Education Department

A. W. JEFFREY, H.M.I., Scottish Education Department (*Secretary*)

Contents

vii

Introduction

This Teachers' Guide corresponds to Book 2 of the Pupils' Textbook of *Science for the Seventies*, and covers Units 9–15 of the Scottish Integrated Science Syllabus.

The Introduction to the *Teachers' Guide to Book 1* (which deals with Units 1–8) gives a full statement of the philosophy behind the course and its objectives. Some recommendations about teaching method are also given, and the role of the *Science Worksheets* published in connection with the course is discussed. It is not necessary to repeat this material here and teachers not already acquainted with it are recommended to read the relevant section of Book 1 (pp. 1–9) before using the present volume.

As with the *Teachers' Guide to Book 1*, the present book is intended to be of use with the relevant *Science Worksheets* (i.e. sections 9–15) as well as with the *Pupils' Textbook*. The latter contains all the experiments covered by the *Science Worksheets* and many others in addition. Cross-references between the experiments in the *Pupils' Textbook* and the *Science Worksheets* are given in Appendix 2 of this guide, so that teachers using the *Science Worksheets*, but not the *Pupils' Textbooks*, will be able easily to locate the information relating to them.

The Teachers' Guide is intended to help the teacher by giving hints on carrying out the experiments, notes on teaching method, lists of equipment required for each experiment, and in cases where difficulty might be experienced, sources of material. It is, of course, based on our own experience and on that of other science teachers with whom we have come in contact, but it can make no pretence at being fully comprehensive, and the authors would welcome comments and additional information which could be included in any future edition of the book.

The authors are indebted to the Controller of Her Majesty's Stationery Office for permission to quote the specific objectives for each unit from *Curriculum Papers 7—Science for General Education*. This publication, which is the Report of the Working Party on Secondary School Science set up by the Scottish Education Department, gives, together with much other very valuable material, a syllabus for an Integrated Science course. It is this syllabus which forms the foundation of the series of books entitled *Science for the Seventies* and

1

the *Science Worksheets*, and the Report is therefore of fundamental importance to all teachers using the course.

We are also grateful to the Scottish Education Department and the Controller of Her Majesty's Stationery Office for permission to reproduce the objective test questions which form Appendix 1 of this guide.

It should be mentioned that the *Science Worksheets* were compiled by some members of the Working Party referred to above on the basis of material collected and tried out in a large number of pilot schools. It is therefore impossible to identify the authorship of the sheets. In order to make the *Pupils' Textbooks* and the *Teachers' Guides* as useful as possible to all concerned with the teaching of the Scottish Integrated Science Course the authors have used the *Science Worksheets* as the core of the course, and they would therefore desire to express their thanks to the large number of science teachers throughout Scotland who contributed to their production.

Unit 9: Making Heat Flow

Specific objectives

The objectives of this unit are that the pupils should acquire:

1. the knowledge that heat energy is transferred in three ways, by conduction, convection, and radiation,
2. further knowledge of the concept of energy,
3. ability to apply this knowledge to new and problem situations,
4. ability to analyse data and draw conclusions (factors affecting heat loss and gain by one of these processes),
5. ability to analyse complex situations to identify the elements (identifying individual methods of heat transfer within a complex),
6. awareness of the phenomena of conduction, convection, and radiation, defined in operational terms,
7. awareness of the importance of heat to mankind,
8. awareness of the need for conservation of sources of heat energy, and
9. skill in the use of measuring instruments and simple apparatus.

Timing This unit will probably form the opening topic for the second year of the course in most Scottish secondary schools. In order to renew interest after a lengthy period of vacation it is desirable that the science topic with which the term opens should be one in which there is a large number of interesting experiments carried out either by the pupils individually or in groups of two. The topic of heat flow, taught in the way we suggest, fulfils this need admirably. Apart from one or two opening demonstrations, the major part of the work consists of a series of 'stations' experiments which gives ample scope for pupil practical work.

Heat flow has usually been included in elementary science courses in the past, and can readily be justified as it not only forms a good introduction to the study of heat, but also has many applications in everyday life. It has normally been taught as a topic in its own right; in the context of the present syllabus, however, there is much more opportunity of presenting it in the light of the particulate nature of matter and the concept of energy, as so much of Book 1 has been concerned with the application of these ideas to the properties of matter. It has already been pointed out that these two concepts form

3

essential 'threads' on which 'integrated' science is based. The idea that heat is a form of energy will be well known to the pupils, and the term 'heat' should be confined to meaning heat energy leaving or entering a body.

It is not necessary to spend much time on the teaching of this unit; whereas in the past it has often been made to stretch out over something like half a term, it is suggested that eight periods should be sufficient to cover all the work, with the pupils doing all the major experiments for themselves.

Apparatus

In the experiments used in this unit, several variations in apparatus have been cited which may perhaps be new to some teachers who have been accustomed to the more stereotyped apparatus. It is not intended to infer that all the familiar apparatus for showing heat transmission is now no longer of use. It will sometimes be possible to substitute it for some of the materials suggested, or it may be used to provide further experiments in the series of 'stations'. Practically all the apparatus required should be already available in some form in schools; that which is not can readily be assembled by a technician or by pupils in a science club.

Particles, energy, and how heat travels

It is suggested that Experiments 9.1, 9.2, and 9.3 be taken as demonstrations by the teacher, with, of course, the pupils co-operating wherever possible. This will introduce the pupils to the ideas of conduction, convection, and radiation, so that when they come to carry out the series of 'stations' experiments they will not be at a loss to interpret their results.

Experiment 9.1 is a familiar one on conduction. It will be necessary to revise the pupils' knowledge of the kinetic theory of matter. We make use of the picture of a solid as composed of particles in a lattice. It is sufficient to refer only to increased vibrational energy of the particles due to their absorption of heat energy. The pupils have already used this idea in predicting the phenomenon of expansion and explaining change of state. To illustrate what takes place in conduction pupils may be placed in a line, shoulder to shoulder. The end pupil should be asked to sway backwards and forwards but not to change his position. Each pupil in the line will in turn take up the vibration. It is recognized that this is not a perfect analogy, but it is better than another which has been commonly used, that of passing a ball from hand to hand down a line of pupils, because it brings out the idea that it is energy that is passed on, and that the particles retain energy because it is being continuously fed in at the end of the line.

Experiment 9.1 (p. 1)
Heating a metal rod

D.

Copper rod.
6 tacks, rivets, or drawing pins.
Wax candle or vaseline.
Retort stand.
Bunsen burner.

The rivet nearest the heat source falls off first. A cardboard or asbestos sheet can be threaded over the rod between the source and the first rivet to screen off radiant heat. The need for this may well be pointed out by the pupils themselves.

Experiment 9.2 (p. 2)
Heating a liquid

D.

Permanganate crystal.
Net cloth.
Beaker (400 cm³).
Bunsen burner, tripod, gauze.

A well-polished beaker of at least 400 cm³ capacity should be filled with cold water to within 2 cm of the brim, the outside of the beaker dried, and then placed on the wire gauze. A fairly large crystal of potassium permanganate should be selected, placed on a wet finger tip and then pushed under the surface of the water close to the edge of the beaker. The crystal should sink slowly to the bottom. A low flame should be applied directly underneath the crystal (say gas about half-way on and air hole open but without a cone showing).

The bunsen burner should be removed as soon as circulation of the 'dye' stream has been established.

To explain convection it will be necessary to mention density, but this should be done qualitatively by referring to increase in spacing between particles when heated.

In Experiment 9.3 we are introducing radiation. As a radiator either the Nuffield Element or a vertically mounted piece of wire gauze heated red hot by a bunsen flame can be used. If using the former, great care must be taken as the element is bare and will be connected to the mains.

A radiant heat detector can be purchased commercially, or can be made cheaply using the following circuit.

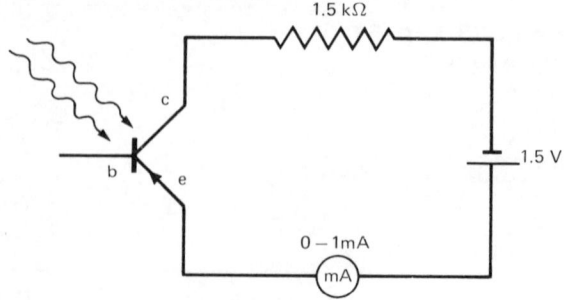

FIGURE 9.1

Any resistor in the neighbourhood of 1.5 kΩ is suitable. The meter could be a cheap Japanese milliammeter costing about £1.30.

The central wire base of the transistor should not be connected. The collector wire is usually marked with a spot and should be connected to the negative terminal of the cell. Various methods of wiring can be used; one neat form consists of mounting the components inside plastic tubing.

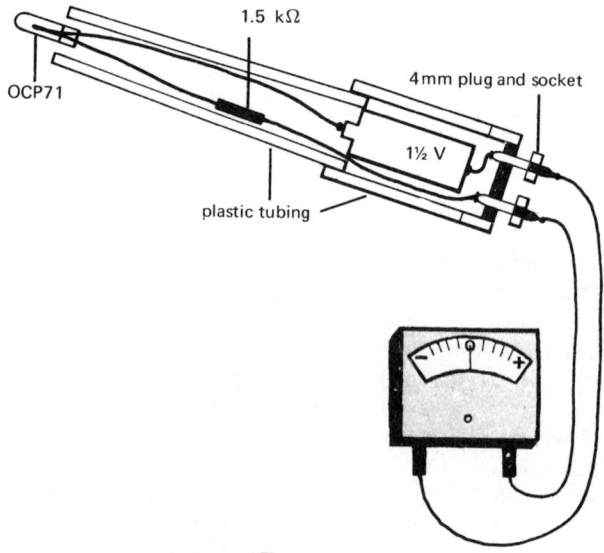

FIGURE 9.2

Experiment 9.3 (p. 2)
Heating by waves
D.

Heat radiator.
Transistor (e.g. OCP 71) with 1.5 kΩ resistor and 1.5 V cell.
Milliammeter (0–1 mA).

More experiments on heat flow

Once the three methods of heat transmission have been established, the pupils may work round a series of experiments laid out for them in the laboratory. The suggested experiments are set out on the *Science Worksheets*, Section 9, 1–6, and are also given in the *Pupils' Textbook*. A card should be placed at each 'station' with brief instructions, or reference to the *Science Worksheets*, or to the *Pupils' Textbook*. The time allocated to each 'station' should be just over ten minutes, five minutes of which is required, in some cases, to record temperature changes. The movement round the 'stations' circuit should go with a swing. Pointing out the applications relevant to the discoveries at each station should be held over to the part of the lesson where the results of all the groups are revised and collated. Stations 8 and 10 take up less time than the others and could be combined into one of the time intervals allocated. Some stations could be duplicated according to the number of pupils in the class and the size of groups.

Station 1. **Experiment 9.4 (p. 3)**
Comparing the rates at which heat is conducted in different solids
S.

Set of metal rods—copper, aluminium, brass, iron, and a hard glass rod, all of the same length and cross-section.
Heat-sensitive paper or matches.
Retort stand.
Bunsen burner.

Rods of copper, aluminium, brass, and iron may be used. The *Science Worksheets* suggest glass as one of the substances. If this is used it should be a hard glass rod; the usual laboratory glass rod, made of soda-glass, may soften when heated. The rods should all have the same length and cross-section. The reason for this should be obtained from the pupils. They should be arranged so that one end of each of the rods can be heated all at the same time. Match heads may be placed at the ends of the rods, or slid along them from the cold end till they reach the point on the rod where they ignite. Alternatively, strips of heat-sensitive paper (cobalt chloride paper) can be placed along the rods to show the rate at which heat is transmitted through the material. Pieces of phosphorus placed on the cold ends of the rods should *not* be used. The order of conductivity, starting with the best, is copper, aluminium, brass, iron, glass.

Station 2.   Experiment 9.5 (p. 3)
Does heat travel through water equally well both upwards and downwards?
S.

2 test-tubes.
2 thermometers (0 to 110° C).
2 retort stands.
2 bunsen burners.

The test-tubes should be dried carefully on the outside. The bunsen burner flames should be adjusted so that the gas taps are turned on half way and the air hole is in such a position that there is no cone.

It is always a temptation for pupils to remove thermometers from a liquid in order to read them, and they should be warned against this.

An alternative method is to include a small piece of ice in tube B, weighted so that it sinks to the bottom. The ice should remain unmelted for the period of each experiment.

Station 3.   Experiment 9.6 (p. 4)
Is air a good conductor of heat?
S.

2 combustion tubes (15 cm).
4 one-holed rubber stoppers to fit.
2 thermometers.
2 retort stands.
2 bunsen burners.

This experiment may be new to some teachers. Again a low flame is advised. It is important to note that the stoppers which are not fitted with thermometers should also have holes in them, as the air in the glass tubes will expand when heated. Thermometers are most easily fitted into one-holed rubber stoppers by selecting a cork borer into which the thermometer will just fit. The borer should then be screwed into the hole, the thermometer pushed through the borer to the required distance, and the cork borer carefully withdrawn with a screwing motion. This will leave the thermometer gripped by the rubber.

Station 4.   Experiment 9.7 (p. 5)
What makes the propeller move?
S.

Bunsen burner.
Foil spiral or propellers

The apparatus for this experiment may take several forms. Paper spirals are apt to catch fire, and so foil, preferably copper, is recommended. Propellers of various kinds made by the pupils can also work well. One very simple type can be made from an aluminium milk-bottle top by making radial cuts, about 0.5 cm long, with scissors from the circumference in towards the centre at 0.5 cm intervals all round. The bottle top can now be bent into a series of blades as on a turbine. A long pin pushed through the centre, and then held vertically acts as a bearing. The pin can also be pushed into the end of a pencil and the propeller can be placed about 25 cm above a lit bunsen burner. It will then rotate vigorously.

The pupils should be asked to blow on the propeller in order to appreciate that it is the hot air rising which is responsible for the rotation. When discussing the implications of this experiment with the class reference should be made to the principles of ventilation. Hot stale air rises, and cold fresh air enters lower down.

Station 5.   Experiment 9.8 (p. 5)
Which cools faster?
S.

2 conical flasks (400 cm³).
2 thermometers.
2 one-holed stoppers to fit.
2 asbestos mats.

The paint for flask A should be matt black. This flask should radiate heat much faster than the polished flask, B. Flask B can alternatively be painted with aluminium paint. Once the concept is clear reference can be made to the use of black in painting pipes at the back of a refrigerator in order to radiate heat, and for air-cooled motor engines, etc. The use of silvered surfaces for minimizing heat loss by radiation, e.g. in metal tea-pots and the like, can be referred to.

Station 6.   Experiment 9.9 (p. 6)
Reflecting heat
S.

2 flasks (400 cm³).

2 one-holed stoppers to fit.
2 thermometers.
1 or 2 100 watt lamps, suitably clamped for attachment to mains point.

Only one 100 watt lamp need be used if the bottoms of the flasks can be positioned equidistantly from the lamp and at the same angle. The temperature rise in flask B should be significantly greater than in A. If the flasks are placed below the lamps it should be obvious to the pupils that the energy cannot be transmitted by convection, as hot air rises. They have also found that air is a poor conductor. Hence the energy must have been transmitted by radiation.

In going through the implications of this experiment with the pupils reference should be made to the fact that this is the method by which energy travels the approximately 93 000 000 miles (150 000 000 km) to us from the sun in about 8 minutes 20 seconds, at 186 000 miles per second (3×10^8 m per second).

The use of silver or bright colours to reflect heat should be referred to. Examples include dairies on farms, sports dress for cricket and tennis, summer clothing, refrigerators, etc. Do not refer directly to the examples listed in the section headed 'Some questions for you' at the end of the chapter in the *Pupils' Textbook*.

Station 7.   Experiment 9.10 (p. 6)
The vacuum flask

S.

2 vacuum flasks, one with broken seal.
2 one-holed stoppers to fit.
2 thermometers ($-10°$ to $110°$ C), the thermometers to have the range $60°$ to $100°$ C visible above the stopper.

This experiment uses two vacuum flasks in one of which the seal is broken so that there is air between the walls of the flask. Breaking the seal without smashing the flask is not an easy matter. It can be done under water with a pair of long-nosed pliers gripping the seal and then twisting the pliers. Water will, of course, enter the walls of the flask and must be shaken out.

The flask with the intact seal should lose practically no heat, while that with the seal broken will lose heat by convection and a little by conduction. In the intact flask heat cannot be lost by conduction, convection, or radiation, nor can it be gained by any of these methods, and so the flask is equally suitable for keeping cold things cold as for keeping hot things hot.

Station 8. Experiment 9.11 (p. 7)
A peculiar light bulb
S.

2 60 watt lamps painted half matt black and half silver.

Two 60 watt light bulbs should be available, painted as shown. A pupil should place the palm of each hand against opposite sides of the bulb. The bulb should not be switched on for much longer than 5 seconds, at the end of which time the heat given out by the black side is quite easily discernible to be greater than that from the silvered side. If the other partner in the group is to repeat the experience the second bulb should be fitted in exchange. For the sake of setting a good example, the circuit should be unplugged from the mains before changing the lamp.

It is suggested that the emphasis should be on the fact that the apparatus shows that dull, black surfaces radiate better than bright, silvery ones but, of course, the black surface is also absorbing more heat from the filament.

Station 9. Experiment 9.12 (p. 7)
S.

Heat source—Nuffield Heating Element, or bunsen burner heating a piece of wire gauze held vertically in a clamp. If the Nuffield element is used a warning should be given about its danger (see Experiment 9.3).
2 sheets of tin plate, one painted black, the other left shiny.
Wax candle or vaseline.
2 tacks.

The metal sheets should be cut from tin plate about 30 cm by 8 cm, with feet cut out and bent as indicated in Figure 9.3.

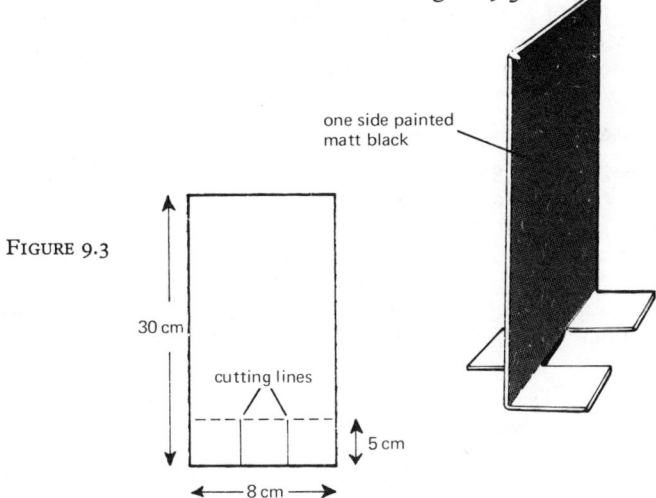

one side painted matt black

FIGURE 9.3

30 cm

cutting lines

5 cm

←— 8 cm —→

One side of each pair of sheets should be painted matt black. A tack should be attached to the back of each sheet with candle grease or wax. The sheets should be placed at equal distances on either side of the heat source. The tack slides from the back of the blackened sheet first.

Station 10.   Experiment 9.13 (p. 7)

S.

Wood-metal cylinder (as in Figure 9.4).
Sheet of paper.
Bunsen burner.

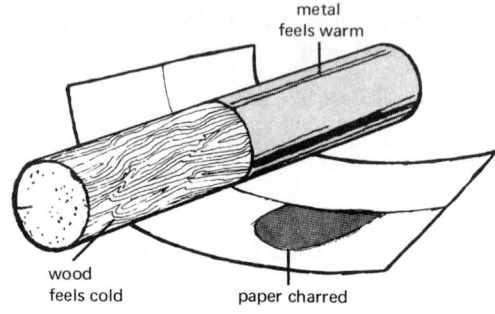

metal
feels warm

wood
feels cold   paper charred

FIGURE 9.4

After the paper has been warmed with a low flame for 5 to 10 seconds it should be found that the paper next to the wooden section has charred, while that next to the metal has not. The wood still feels cold, while the metal is now warm. Obviously the wood has not conducted the heat away from the paper which has consequently become charred; the metal, on the other hand, takes the heat away from the paper, which therefore does not char.

Pupils should be helped to realize that the sensation of cold that they experience is due to heat flowing away from them. It could be pointed out that it is possible to be burned by touching a very cold object, the rate of heat flow away from their body being fast enough to result in a burn.

How animals keep warm

Experiment 9.14 (p. 8)
Hair, fur or feathers?

P. 4 groups.

4 thermometers.

4 flasks (500 cm³) insulated, one with cotton wool, one with fur, and one with feathers, while one is uncovered.

4 one-holed stoppers to fit, to take thermometers.
Graph paper.
Stop clock.

The following experiment can be done as a group experiment. Four sets of pupils will be required to note the temperatures of the water at set intervals in each of the four flasks.

The different materials to cover each flask are best held on by elastic bands.

The flasks should be almost completely filled with nearly boiling water, and the part of the thermometer stems from 50° to 100° C should be above the level of the stopper.

The temperature of the water in the uncovered flask should fall much faster than that of the others.

The pupils should realize that the insulating effect of the air trapped in the cotton wool, fur, and feathers is important.

Experiment 9.15 (p. 9)
Which is the best for keeping things hot?

1, 2, or 3 sets of cans as described below.
4, 8, or 12 thermometers.

As with Experiment 9.14 this may be done as a group experiment, or in sets of 8 pupils. The rate of cooling of four vessels over 10 minutes has to be found and the class can be divided up accordingly.

A small empty soup tin is suitable for can A, and C can be a similar can placed inside a larger one with lagging of expanded polystyrene cut from packing material. B can be a picnic cup, and D should be a polystyrene cup provided with a lid. The thermometers should be supported in clamps to avoid their falling out of the cans and breaking.

Comparing A with C, C will lose less heat than A by conduction and convection.

Comparing A with B, B will lose less heat than A by conduction.

Comparing B with D, D will lose less heat than B by convection because of its lid.

D should retain its heat best.

In the *Pupils' Textbook* a list of problems is given under the heading 'Some questions for you'. These may be given for homework, or dealt with as revision points in class.

Further applications

Question 9 in the 'Some questions for you' section deals with the domestic hot-water system and with land and sea breezes. Most schools have a commercially produced model of a domestic hot-water system, but satisfactory ones can often be made by members of the science club, or by technicians. It is best to use a thermochromic liquid (a liquid which changes colour with temperature) in the model, if obtainable. Otherwise crystals of potassium permanganate can be put in. After a time, however, potassium permanganate gives a brown deposit of hydrated manganese(IV) oxide and it is advisable to remove the potassium permanganate solution each time after the model has been used. The stain produced by hydrated manganese(IV) oxide can be removed by adding a drop or two of concentrated hydrochloric acid. The pupils should be asked to trace out the circulation of hot water between the boiler and the storage tank, and to find the hottest part of the latter (the top).

Land and sea breezes provide useful examples of convection in nature. If pupils are assisted to work out the direction of the breeze during the day—from the sea to the land—they should be able to work out for themselves the direction at night.

Useful visual aids for this unit

Film loops (Casettes for 800 E)

SP/H/2   Emission and absorption of radiation.
SP/H/3   Reflection of radiation.
SP/H/4   Convection of water.

All the above are obtainable from Ealing Scientific Ltd., 23 Leman Street, London, E.1.

16mm Films

D 2184   Heat—its nature and transfer.
DB 6   Leslie's cube.
DCF 2664   Convection.
DCF 2665   Conduction.
DCF 2666   Radiation.

All the above are obtainable on loan from the Scottish Central Film Library, 16/17 Woodside Terrace, Glasgow, C.3.

The ESSO film 'Experiments in Heat Radiation' may be loaned from Travelling Films Ltd., 78 Victoria Road, Surbiton, Surrey.

Unit 10: Hydrogen, Acids and Alkalis

Specific objectives

The objectives of this unit are that the pupils should acquire:

1. knowledge of the properties of hydrogen,
2. knowledge that water is formed when hydrogen burns,
3. knowledge that certain metals will react with water at room temperature and that others will react with steam,
4. knowledge that certain metals will displace hydrogen from some dilute acids and that others will not,
5. knowledge that there is a gradation of reactivity among the common metals,
6. knowledge that pH is a measure of the degree of acidity of a solution,
7. knowledge that acid and alkali are the names given to solutions at opposite ends of the pH scale,
8. knowledge that acids neutralize alkalis,
9. knowledge that there is a simple quantitative relationship in the neutralization of acids by alkalis,
10. the ability to draw conclusions from a mass of data,
11. the ability to design experiments to test hypotheses,
12. awareness of the processes involved in identifying a chemical substance,
13. awareness of the use of standard scales for comparison purposes,
14. skills in handling simple chemicals and glassware, and
15. awareness of the dangers of handling hydrogen in large quantities.

It will be noticed that the majority of the above objectives involve the acquisition of knowledge. The teacher must, however, avoid dealing with this material in the old traditional way. If the guide-lines laid down in the *Pupils' Textbook* are followed there will be no danger of this happening.

The suggested *time allocation* for this unit is twelve periods.

Order of development

The order of development of this unit in the *Pupils' Textbook* is as follows:

15

1. Pupils discover for themselves the properties of hydrogen, e.g. insolubility in water, flammability, non-supporter of combustion, density, explosion with oxygen.
2. Study of the burning of hydrogen leading to the synthesis of water.
3. The preparation of anhydrous copper(II) sulphate for use as a test for water-containing substances leads to a brief study of water of crystallization.
4. The analysis of water by electrolysis.
5. The action of metals on water and steam leading to an activity series.
6. The reduction of copper oxide by hydrogen, and the reduction of oxides in general as related to the activity series deduced in 5.
7. Tests for acids; introduction of pH as a scale of acidity.
8. The action of acids on metals; further confirmation of activity series.
9. Alkalis; their general properties and action on acids.
10. Application of neutralization; acidity of the soil.

Note on sources of hydrogen

Throughout this course little emphasis has been placed on the preparation of gases *per se*. Whereas in the traditional treatment of inorganic chemistry it was customary first to prepare a gas and then carry out tests for its properties, the approach here is to regard the gas under study as if it were any other chemical and get it 'from the bottle'. This means that pupils must be provided with tubes of gas with which to experiment, and involves the teacher or his technician in a good deal of preparation prior to the lesson.

Cylinders Although it is convenient to obtain the gas for pupil experiments from cylinders it must be understood that these are potentially dangerous—even more so than cylinders of other gases. Because of its low density, hydrogen is particularly likely to leak and cylinders must be stored away from any flames. In schools they should be kept in a special store outside the building. If it is necessary to move them this should only be done using a proper trolley for the purpose. When jars or test-tubes are being filled from a cylinder (or from an aspirator) there should be no flames in the room. Not only flames, but electric sparks can explode a mixture of hydrogen and air, and accidents are known to have been initiated by such simple events as switching off an electric light. Throughout the work in this unit, *hydrogen must be treated with the greatest caution*. It is safe to hold a small test-tube of the gas near a flame, but larger quantities should not be ignited.

Other methods As the dangers associated with cylinders of hydrogen are so great it is probably wise not to use them at all, but to make the hydrogen as required. This can be done by the action of zinc on dilute hydrochloric or sulphuric acid and storing the gas in an aspirator (Figure 10.1).

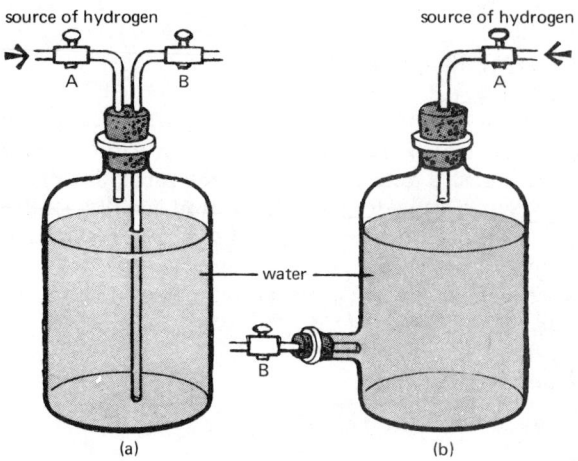

FIGURE 10.1

The gas generating apparatus should be simply a flask or bottle with no thistle funnel. If the usual apparatus with a thistle funnel is employed air is sometimes drawn in through the funnel because of the pressure difference, and this may give a dangerously explosive gas mixture. The clips A and B are open while hydrogen is being passed into the bottle. The bottle must be *completely* full of water before hydrogen is passed in. To drive hydrogen out of the container tube B is connected to the water tap. There must be no air in tube B or in the tubing connecting B to the water tap.

Test-tubes can be filled with hydrogen either by displacement of water or by holding the tube vertically and passing hydrogen upwards into it. With the latter method one can only guess when the tube is full.

Some facts about hydrogen

Experiment 10.1 (p. 11)
Does hydrogen dissolve in water?

P. 10 groups.

10 test-tubes of hydrogen.
10 troughs, basins, or large evaporating dishes.

Given a test-tube full of hydrogen and a basin of water pupils should be able to work out how to do this for themselves. All that is necessary is to shake the tube mouth down in water.

Experiment 10.2 (p. 12)
Does hydrogen burn?

P. 10 groups.

20 test-tubes of hydrogen.

10 bunsen burners.

If a tube is full of hydrogen the gas will burn with a very slight pop when a flame is placed near the mouth of the tube. The flame is almost invisible. A mixture of hydrogen with air burns with a louder pop or a squeal. The object of the second part of the experiment is to obtain a mixture of hydrogen and air.

This experiment should be done with the normal size of test-tube—not a gas jar.

More about hydrogen

Experiment 10.3 (p. 12)

P. 10 groups.

20 test-tubes of hydrogen.

10 tapers.

Bunsen burners.

The object of Experiment 10.3 is to show that hydrogen is less dense than air.

Pupils should complete the sentence on p. 12 of the *Pupils' Textbook* with the words 'the smallest density'.

What is formed when hydrogen burns?

Experiment 10.4 (p. 12)

P. 10 groups.

10 test-tubes of *dry* hydrogen.

Demonstration experiment

Apparatus shown in Figure 10.2.

Thermometer.

Small test-tube (for finding freezing point and boiling point).

Ice and salt.

Bunsen burner.

It is not easy to show that water is formed when hydrogen burns. We start by giving the pupils a test-tube full of dry hydrogen and asking them to see what happens when it burns. They should find that the sides of the tube become covered with a film of liquid. Of course, for this experiment the hydrogen cannot be collected over water.

In order to prove that the liquid formed is water it is necessary to collect more of it so that adequate tests can be carried out. The burning of hydrogen itself at a jet is fraught with danger. If the teacher decides to carry out this experiment *it should be done behind a safety screen*, and he should wear plastic goggles. The hydrogen must, of course, be dry. If it is obtained from a cylinder it can be assumed that this is so, but if it is obtained from an aspirator, or direct by the action of dilute hydrochloric acid on zinc, it must be dried by passing through a tube containing a drying agent such as silica gel. Suitable apparatus is shown in Figure 10.2.

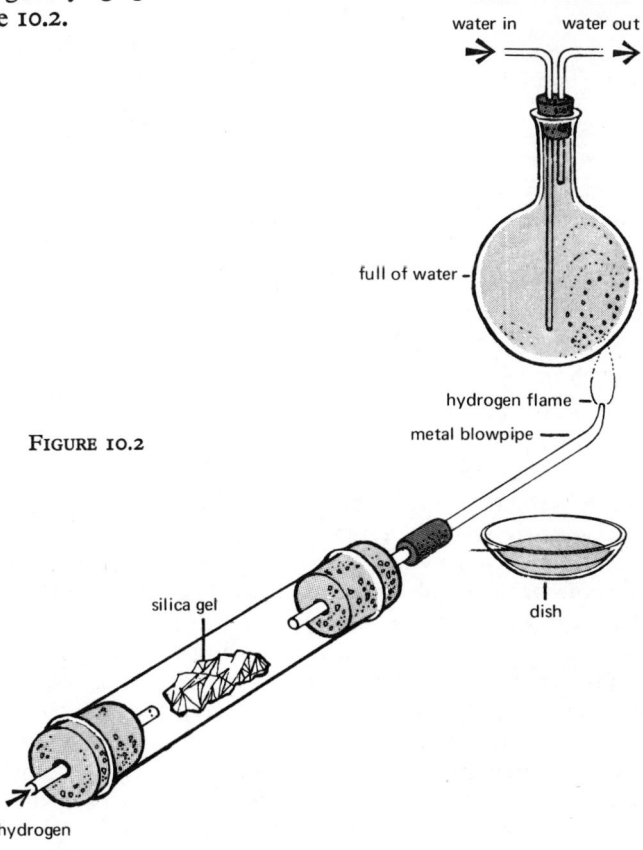

FIGURE 10.2

A very much safer method, but one which is not so valuable educationally, is to tell the pupils that town gas contains hydrogen, and that you propose to use it instead of hydrogen. Strictly speaking the pupils could rightly demand that you prove to them that town gas indeed does contain hydrogen. It would be difficult to do this without assuming what we are trying to prove. The town gas should be dried by passing it through silica gel, and may be burnt at a metal jet. An old metal blowpipe is suitable. The jet should be made to play against a cold surface, e.g. a large flask full of water.

It is unfortunate that there is no chemical test for water that could be applied to a minute quantity of the liquid. All the common tests, such as the turning of anhydrous copper(II) sulphate blue, the effect on cobalt chloride etc., will show only whether or not a liquid *contains* water. Hence we have to fall back on physical properties, and to determine these a fair quantity of liquid is required. Strictly speaking it is necessary to determine at least the freezing point and the boiling point of the liquid; even then some liquid other than water might possess both these properties in common with water. Determination of the density of the liquid is another property that might be added. There are micromethods of determining these properties, but they are not suitable for class demonstration. It will therefore be necessary to collect 1–2 cm^3 of the liquid, and this takes some time. The experiment should be started at the beginning of a period, and the pupils can be engaged with other work while it is proceeding.

Owing to the difficulties associated with this experiment, it is advisable to put more stress on the analysis of water than on its synthesis (see p. 21).

Water in crystals

The purpose of Experiment 10.5 is to introduce the anhydrous copper(II) sulphate test for water in a liquid.

The fact that water of crystallization is combined water should not be difficult for the pupils to comprehend. They know that combination involves drastic changes in the properties of the substances combined (they have come across this in Unit 4, and again in the experiment they have just done on the combination of hydrogen and oxygen to form water). They will discover that when water is added to the anhydrous copper(II) sulphate a good deal of heat energy is given out—again a sign of combination.

As mentioned above the anhydrous copper(II) sulphate test only shows, of course, that a liquid *contains* water, not that it *is* water.

Experiment 10.5 (p. 13)
Where does the water come from?

P. 10 groups.

10 small test-tubes or crucibles.
Copper(II) sulphate (small crystals).
Bunsen burners.

If it is decided to attempt to show that the liquid driven off is water, the teacher should heat some copper(II) sulphate crystals in a distillation apparatus. A hard glass flask is necessary.

Experiment 10.6 (p. 14)
Do all crystals contain water?

P. 10 groups.

10 small test-tubes.
Bunsen burners.
Samples of *sodium carbonate (hydrated), sodium chloride, *magnesium sulphate, potassium nitrate, potassium chloride, *sodium sulphate, potassium sulphate, *cobalt chloride.

Different groups can each use one substance and the results of the class can be collected. Those substances marked with a * contain water of crystallization.

Synthesis and analysis

In Unit 4 pupils came across the making of a compound, and its breaking up. They used copper(II) chloride for this purpose, and they decomposed it by electrolysis. They should be able to suggest that possibly water might be broken up by electrolysis also.

We first discover that water is a very poor conductor and that very little, if any decomposition occurs with the sources of supply available.

To obtain a reasonable quantity of hydrogen and oxygen in the time available we have to add an electrolyte to the water. If we are to be able to conclude anything from the experiment it is obvious that this electrolyte must not contain hydrogen or oxygen. A satisfactory substance is sodium fluoride.

Of course, the process taking place is much more complicated than the mere statement that water is broken down by electrolysis into hydrogen and oxygen would lead us to assume. However, we need not at this stage enter into the mechanics of the process. It is sufficient to say that it is possible to show that the sodium fluoride is still present at the end of the experiment and that none of it has been used up. The pupils might be expected to be able to suggest how to prove this.

Any type of electrolysis apparatus may be used for this experiment (see Figure 10.3). Carbon rods from old dry cells form suitable electrodes.

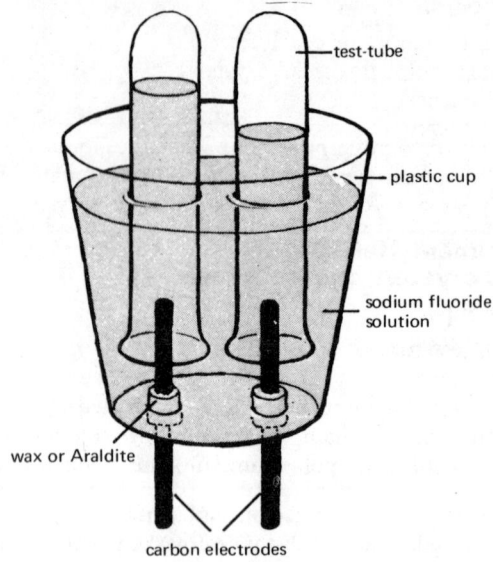

FIGURE 10.3

The pupils should note the approximate relative volumes of hydrogen and oxygen collected, and should be able to devise methods of testing for these gases.

Experiment 10.7 (p. 14)
Breaking up water

P. 10 groups.

10 sets of electrolysis apparatus (Figure 10.3).
10 ammeters.
10 sources of low voltage (power packs, accumulators, or dry cells).
   6 V will be required.
Sodium fluoride.
Splints (for testing for oxygen).
Bunsen burners.

Some peculiar metals

We now pass to other ways of getting hydrogen from water. The metals sodium, potassium, and calcium (and to a slight extent, magnesium) will decompose cold water, they are followed in the electrochemical series by others which will not decompose cold water, but will decompose steam, and then, at the bottom end of the series by those which have no effect on water at all. The pupils will, of course, discover these facts for themselves, and will draw up an activity series of the metals based on their observations.

The action of calcium on water is a perfectly safe experiment for pupils to carry out. The action of sodium on water should only be demonstrated, and care should be taken in case any of the metal spurts from the trough, as it is prone to do. Reference could be made to the reason why sodium is stored under oil. The action of potassium on water is extremely vigorous, and should not be attempted in front of a class.

Experiment 10.8 (p. 15)
Calcium and water

P. 10 groups.

10 test-tubes.
Freshly scraped calcium turnings.
10 filter funnels.
Filter papers.
Glass tubes, or drinking straws, for breathing into the calcium hydroxide solution.
Indicator paper.

Experiment 10.9 (p. 15)
Sodium and water

D.

Trough.
Small pieces of sodium.
Indicator paper.

The effect of water on magnesium under different conditions can be tried. Magnesium powder reacts very slowly indeed with cold water, and it will be necessary to leave the experiment set up for a day or two in order to see if there is any result. Magnesium powder tends to float on water because of surface tension effects; it is therefore suggested that it be wrapped in a small piece of tissue paper. The experiment should be tried also with magnesium ribbon. It should be found that the action on the latter is much slower than on the powder. The explanation of this can be given in terms of kinetic theory, the powder offers a much larger surface for attack than does the ribbon.

If magnesium powder is heated with water, hydrogen is evolved fairly rapidly. Again, some explanation can be given in terms of kinetic theory.

The 'wet asbestos' method can be used to try the effect of heating metals in steam. Instead of using asbestos, however, which has been shown to be carcinogenic, we substitute 'rocksil', a material used for insulating purposes in building. It can be obtained from builders' merchants. The apparatus is shown in P.T. Figure 10.4.

Magnesium powder must not be used in this experiment as the reaction is too vigorous, and explosions have occurred.

The method can be used with other metals (Experiment 10.11). The vigour of the reaction can be gauged by the effect on the tube. With magnesium the heat evolved is sufficient to melt the tube. With zinc, use zinc foil, not powdered zinc.

Experiment 10.10 (p. 15)
Magnesium and water

P. 10 groups.

Magnesium powder.
Magnesium ribbon.
20 test-tubes.
Rocksil.
10 retort stands.
Bunsen burners.

Experiment 10.11 (p. 16)
Other metals and water

P. 10 groups.

Samples of zinc foil, iron filings, tin foil, lead foil, copper foil (or powder).
10 test-tubes.
Rocksil.
10 retort stands.
Bunsen burners.
Different groups should use different metals and the results for the class can be pooled.

A chemical tug-of-war

We now want to draw some predictions from the activity series. The argument could run like this. We have seen that copper is unable to turn out hydrogen from its compound with oxygen under the conditions we have used. Hence hydrogen must have a greater affinity for (or 'pull on' or some such phrase) oxygen than copper has. What then would we predict would happen if we were to pass hydrogen over copper oxide? Let's try it and see if we are right. Can you suggest what apparatus to use?

It is useful to see if anything happens when the reagents are cold. (The experiment has to be started this way anyhow, but few teachers have ever drawn the pupils' attention to the lack of reaction in the cold.)

The experiment is *dangerous*, and, if carried out with hydrogen, must be done behind a safety screen. As with Experiment 10.4, however, town gas can be used in place of hydrogen. If the pupils have accepted this in the earlier experiment there should be no difficulty in using it again here. Even using town gas there are dangers. It is essential to make certain that all the air has been swept out of the apparatus before heating the copper oxide.

Butane does not reduce copper(II) oxide at temperatures normally attainable in the laboratory.

It is suggested that pupils should not simply accept the evidence of colour for the fact that reduction has taken place, but should test the product with nitric acid. The presence of metallic copper is shown by the production of the brown gas, nitrogen dioxide.

It will be seen that a 'word equation' has been introduced in the *Pupils' Textbook* in section 10.8; it is expected that at the end of Experiment 10.12 the pupils should be able to write a similar equation for the reaction that has occurred.

Experiment 10.12 (p. 17)
Hydrogen vs. copper
D.

Apparatus as shown in P.T. Figure 10.6.
Dry copper(II) oxide.
Source of hydrogen or town gas. The gas should be dried by pass
ing over silica gel.
Nitric acid (diluted with its own volume of water).

Acids

The use of pH paper to test whether a liquid is acidic is introduced in Experiment 10.13. It is possible to use litmus paper, but it is suggested that universal indicator paper be used as this introduces the pupils to the idea of a scale of acidity. It can be purchased in books or in rolls, but the latter is the more economical. There should, of course, be no attempt to explain the basis of the pH scale.

Experiment 10.13 (p. 18)
Testing for acids
P. I.

A selection of common acids such as vinegar, lemon juice, grape-fruit juice, tartaric acid, citric acid, sour milk.
20 test-tubes or watch glasses.
Indicator paper (p. 18).

Experiment 10.14 (p. 18)
Testing a solution of carbon dioxide in water
P. I.

Solution of carbon dioxide in water.
20 test-tubes or watch glasses.
Indicator paper.

Experiment 10.15 (p. 18)
Does water play a part in making substances acidic?
P. I.

Solid oxalic acid, tartaric acid, citric acid.
20 watch glasses.
Indicator paper.

Metals and acids

We now come to a series of experiments on the action of metals on dilute acids. The pupils should find that a similar activity series is obtained as with the action of metals on steam.

Experiment 10.16 (p. 18)
Will magnesium turn out hydrogen from dilute hydrochloric acid?
P. 10 groups.

Dilute hydrochloric acid (about 4 M).
Magnesium ribbon.
10 test-tubes.
Tapers (for testing for hydrogen).

The solution remaining after the reaction should be retained for Experiment 10.19.

Experiment 10.17 (p. 18)
The action of metals on dilute hydrochloric acid

P. 10 groups.

Dilute hydrochloric acid.
Aluminium foil, iron filings, lead shot, tin foil, copper foil, silver foil (if available), mercury.
10 test-tubes
Tapers (for testing for hydrogen).

Pupils should be warned not to allow mercury to run down the sink. All tubes containing mercury should be collected by the teacher. The metal can be separated by pouring it through a filter paper with a small hole in it, and collecting it in a dry tube. The filter paper will absorb the liquid. Silver foil should also be collected and dried for further use. It is not necessary for each group to try all the metals, but the results of the class can be pooled.

Experiment 10.18 (p. 19)
Metals and dilute sulphuric acid

P. 10 groups.

As for Experiment 10.17, but substitute dilute sulphuric acid (about 4 M) for dilute hydrochloric acid.

Experiment 10.19 (p. 19)
What happens to the metal when it reacts with an acid?

P. 10 groups.

10 water baths (400 cm³ beaker, or can).
10 clock glasses.
Bunsen burners, tripods, and gauzes.

By evaporating the solution remaining after magnesium has reacted with dilute hydrochloric acid the pupils discover that it does not look like magnesium. Discussion leads to the fact that the metal has combined with part of the acid, and a word equation can be drawn up to express what has happened.

The term 'salt' is introduced.

The evaporation is best carried out on a water bath to avoid spirting.

Alkalis

We now introduce the term 'alkali' as meaning something which is the opposite of an acid, in that it has a pH on the other side of 7.

Experiment 10.20 (p. 20)
More tests

P. I.

20 watch glasses.
Indicator paper.
Samples of sodium hydroxide solution (very dilute), sodium car-
bonate solution, lime water, sodium bicarbonate, ammonia.

The idea that an acid 'cancels-out' an alkali is introduced through
the pupils' own observations. The term 'neutralizes' can be used if the
pupils understand the meaning of the word.

Acid is added to a known volume of a solution of sodium hydroxide
from a graduated syringe, and the effect on the pH after the addition
of each cubic centimetre is noted. The teacher should arrange before
the experiment that the solutions are of approximately the same
molarity.

Experiment 10.21 (p. 20)
What happens when acid is mixed with alkali?

P. 10 groups.

Sodium hydroxide solution (about 4 g NaOH per litre).
Hydrochloric acid (about 10 cm³ of concentrated acid diluted to
1 litre).
10 measuring cylinders.
10 graduated syringes.
10 beakers (100 cm³).
Indicator solution.

What is produced when an acid neutralizes an alkali?

The pupils discover that the product of the neutralization of an acid
by an alkali is a salt. It is suggested that it is safe to identify the product
in the case of sodium hydroxide and hydrochloric acid by tasting it,
however, the teacher *must* warn the class that this is not a normal
procedure in the laboratory, and that it is dangerous to taste things
unless specifically told to do so. Moreover this is not a good test for
sodium chloride as there may well be many other substances that taste
the same as salt.

Experiment 10.22 (p. 21)

P. 10 groups.

As for Experiment 10.21 with indicator paper in place of indicator solution.

The next step is often a rather difficult one to get across. We want
the pupils to realize that there is a quantitative relationship between
the masses of acid and alkali which react, and that as far as this relation-

ship is concerned the quantity of water present has no effect. The pupils find out first of all that equal volumes of solutions of alkali of different concentrations require different volumes of acid of a given concentration for neutralization. By making one alkali solution half as concentrated as the other the pupils are able to see that there is a simple relationship between concentration and the volume of acid required. In Experiment 10.24 they take solutions of alkali which contain the same mass of solute in different volumes of water, and find that the volume of acid required is the same each time. Hence the volume of water added makes no difference; the volume of acid required for neutralization depends only on the mass of alkali present.

Experiment 10.23 (p. 21)

P. 10 groups.

Sodium hydroxide solution (approximately 1 g of solid NaOH in 20 cm³ of solution).
Hydrochloric acid (about 100 cm³ of concentrated acid in 1 litre of solution).
10 graduated syringes.
Indicator solution.

Experiment 10.24 (p. 21)

P. 10 groups.
As for Experiment 10.23.

Applications of neutralization

For Experiment 10.25, pupils should bring in as many samples of widely different types of soil as possible. The method has been left for the pupils to devise. The following points might be discussed. Should equal weights of soil be taken, and the same volume of water added to each? If so, should the soil be dried first? Would it be good enough just to insert a piece of indicator paper into the soil?

In testing the pH of soil the volume of water used is not critical. However, it is necessary to use distilled water, and the tubes used should be thoroughly cleaned. It has been found that new test-tubes often give a marked alkaline reaction to distilled water.

Drying the soil may well decompose some of the acids present. What the agriculturist wants to know is the pH of the soil under the conditions in which he wishes to grow crops. This is clearly not in dry conditions, nor in the presence of large volumes of water.

Experiment 10.25 (p. 22)
Testing soil for acidity

P. 10 groups.

Samples of soil.
10 test-tubes.
Indicator paper.

A practical test

1. Which of the given liquids could possibly be water?
2. Make a few crystals of sodium nitrate from sodium hydroxide solution and dilute nitric acid.
3. Is hydrogen given off when the metal is added to dilute hydrochloric acid?
4. Which of the two soil extracts is the more acidic?
5. What volume of the hydrochloric acid would react with 1 g of sodium hydroxide?
6. Find out if magnesium oxide dissolves in water by using indicator paper.

Notes on the test

1. Provide sodium carbonate solution, very dilute hydrochloric or sulphuric acid, alcohol (colourless, not the purple methylated spirit), and water itself. The pupils should use indicator paper to find that two of these are neutral. One of the neutral substances can be detected by its smell, hence the other must be water.
2. Provide dilute sodium hydroxide solution and dilute nitric acid of approximately equal molarities. The pupils should evaporate a small quantity of their solutions on a water bath, or under an infra-red lamp.
3. It is best to provide a metal they have not used in the course—say an alloy of some kind, such as brass.
4. The pupils can be given either the soils themselves, or so-called extracts which could be made up artificially.
5. This involves some calculation which only the more able pupils may be able to do. Provide a solution of sodium hydroxide of known concentration, say 5 g per 100 cm^3, so as to make the calculation simple. (The concentration can, of course, be fictitious.)
6. Magnesium oxide gives an alkaline reaction with water. It must therefore be soluble to a certain extent, although it may appear to be insoluble.

Unit 11: Detecting the Environment

Specific objectives

The objectives of this unit are that pupils should acquire:

1. knowledge of some facts about the human eye,
2. knowledge of some facts about a pinhole camera,
3. knowledge that the focal distance of a lens is related to its curvature,
4. knowledge of some facts about a lens camera,
5. knowledge that the brain does not always interpret the signal from the eye correctly,
6. knowledge of the major parts of the ear (drum, bones, inner ear),
7. knowledge of the operation of the bones of the inner ear,
8. knowledge that the production of sound requires a vibration,
9. knowledge that pitch is related to frequency, which is related to length of vibrator, and tension in vibrator,
10. knowledge that a medium is needed for transmission of sound,
11. knowledge that the ear has a limited band of reception,
12. knowledge of some facts about taste and smell,
13. knowledge that touch nerve endings vary in concentration in different parts of the body,
14. knowledge of reflex action in muscle/nerve systems and the fact that this reflex takes time to act,
15. ability to make comparisons between related entities (eye and camera),
16. ability to use inductive processes of thought to build up the hypothesis that vibrations are necessary for sound to be produced,
17. ability to design experiments to investigate stated hypotheses,
18. ability to draw conclusions from a variety of data obtained in finding threshold frequencies for the ear,
19. ability to deal with problems with several variables using the effects of smell and feel on taste,
20. awareness of the importance of knowing that the brain may not interpret signals from the eye reliably,
21. awareness of our reliance on binocular vision for many judgments,
22. awareness of the receptors of communication and man's dependence upon them,

23. awareness of the limitations of taste, smell, and touch,
24. awareness of the different levels of control man has over his own muscle structure,
25. awareness of the need for instruments to overcome man's limitations and the inevitable limitations of instruments as well, and
26. some skill in the use of dissecting instruments.

The suggested *time allocation* for this unit is twenty-four periods.

Order of development

One of the basic characteristics of living things is their ability to receive and react to stimuli from their environment. The purpose of this unit is to introduce the pupil to the receptor mechanisms in his own body and to their limitations. The physics of optics and sound is dealt with in relation to the biology of seeing and hearing. The order of treatment in the *Pupils' Textbook* is:

1. The eye and light; the lens of the eye compared with a glass lens.
2. Vision: colour vision, the blind spot, the functions of binocular vision.
3. The ear and sound; structure of the ear; transmission of sound; pitch and frequency.
4. Taste, smell, and other senses.
5. The nervous system. Reflex action.

The eye and light

If the teacher has not had experience of cutting open a bullock's eye, it is essential that this be practiced in advance of the lesson. Instructions are given in the *Pupils' Textbook*, p. 24. It is usually necessary to give the nearest slaughter house or butcher some days' notice in order to obtain the necessary materials on time. After any work involving dissection or the handling of dissected material the teacher should insist that the pupils wash their hands before leaving the laboratory. Dissecting instruments should also be thoroughly washed in detergent.

Experiment 11.1 (p. 24)
Dissecting a bullock's eye

(This experiment may be done as a demonstration, or by groups of pupils depending upon the quantity of materials available.)

Bullocks eyes.
Razor blades.
Scalpel or scissors.

Dissecting tray.
Pins.
Sheets of paper.

The pin-hole camera

The construction of a pin-hole camera is detailed in the *Pupils' Textbook*.
An alternative method is shown in Figure 11.1, where, by cutting out
stray light falling on the screen the image is more distinctly seen.

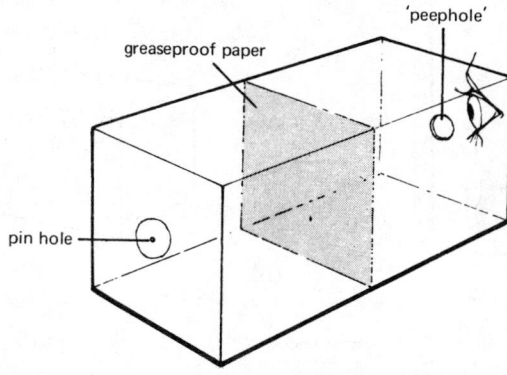

FIGURE 11.1

Experiment 11.2 (p. 25)
Making a pin-hole camera

P. 10 groups.

Cardboard boxes.
Grease proof paper.
Scissors.
Sellotape.
Copper foil.

The image in a pin-hole camera

It is not intended to make this a section on geometric optics. The
pupils are going to use ray boxes shortly and it could then be demonstrated that light travels in straight lines, or, if desired the point
can easily be made here by considering the nature of shadows. One
might then go on to discuss how a candle flame gives light. If a porcelain dish is held over a candle flame soot will be deposited on it;
the flame can be regarded as a collection of glowing carbon particles
each sending out light in all directions. The pupils should be asked to

trace out which rays in Figure 11.4 of the *Pupils' Textbook* (p. 25) will manage to pass through the pinhole. The rays from the top of the flame will reach the bottom of the screen and should explain the inversion of the image.

Alternative light sources are shown in Figure 11.2.

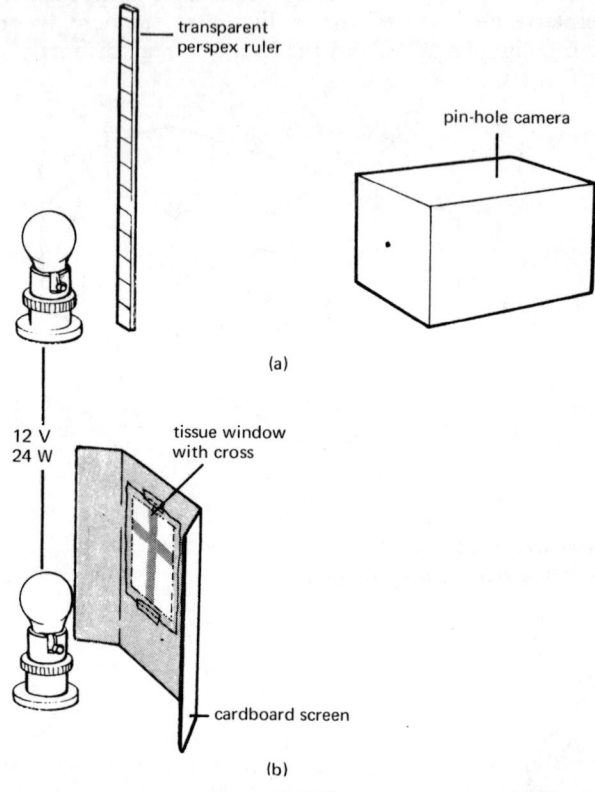

(a)

(b)

FIGURE 11.2

In (a) a transparent perspex ruler is illuminated from behind by a car headlamp. In (b) a window in a cardboard screen is illuminated.

The abler pupils will realize that not only is the image inverted but it is also reversed.

Experiment 11.2 (a)

P. 10 groups.

| Pin-hole cameras which pupils have just made.
| Light sources, such as a candle in a holder, or as in Figure 11.2. |

The study of the pin-hole camera continues with the investigation of what happens when a convex lens is placed between the light source and the pin hole. It is important that the focal length of the lens issued to each group should suit the dimensions of the box of the camera. As a rough guide the focal length of the lens should be rather less than the distance between the pin hole and the screen—say about three-quarters of this distance.

Experiment 11.3 (p. 26)

P. 10 groups.

Pin-hole cameras which pupils have just made.

10 convex lenses (f about three-quarters of distance between pin hole and screen).

The action of the lens

To investigate the action of a lens it is simplest to use a ray box. There are several types on the market but they are not difficult to construct in the workshop from a prototype. The ray box normally uses a 12 volt car head-lamp as a light source. Some low voltage power supply will be needed for this. A lab-pack is recommended. It does not matter, of course, whether a.c. or d.c. is used to light the lamp. A prism is used to introduce the way in which a lens functions. When the pupils have seen that a prism causes divergent rays to converge it may be replaced by a half spherical convex lens.

Experiment 11.4 (p. 26)
How a lens works

P. 10 groups.

10 ray boxes with suitable power supplies.
10 prisms.
10 half-spherical convex lenses.

In order to show how the focal length of the lens in the eye is altered it is convenient to use a 'variable-focus eye'. The Skelton model, obtainable as item 125 in the Nuffield Physics list of apparatus, is suitable. This is essentially a cylindrical lens of jelly enclosed by thin sheets of flexible perspex. The thickness of the lens can be changed by adjusting a screw. The jelly is made by warming together the following ingredients in a hot water bath (the parts are by volume): 13 parts glycerine, 10 parts water, 3 parts gelatine, and 2 parts cane sugar. The apparatus could be made fairly easily by technicians or science club pupils.

The rays should be adjusted to fall on the 'retina' of the 'eye'. If now the screw is adjusted, the shape of the lens will be changed, as will its power. When the lens is adjusted to be fat, the ray box should be placed close to the eye so that the rays converge on the retina. If the ray box is now moved further back it will be found that the curvature of the lens has to be decreased and the lens made thinner to re-focus the rays on the retina.

Experiment 11.5 (p. 27)
The jelly lens
D.

Ray box and power supply.
Skelton model eye.

A model eye can easily be made with existing resources. A fairly large spherical flask and three convex lenses are required. The latter have to be fixed to the outside of the flask by Araldite. The focal lengths of these lenses should be:

   (i) exactly the diameter of the flask round the middle,
  (ii) about two-thirds of the diameter, to illustrate short sight, and
 (iii) about $1\frac{1}{3}$ times the diameter, to illustrate long sight.

The light beam shone on the 'eye' should be collimated, i.e. be made parallel. The pupils find it interesting if suitable lenses are available to correct for the 'defect' in each case, i.e., a suitable concave lens to correct for short sight, and a suitable convex lens for long sight.

Experiment 11.6 (p. 27)
P. 10 groups.

10 model eye—flask type.
Fluorescein.
Convex and concave lenses to show correction of defects.

An out of date folding camera is ideal for Experiment 11.7, being generally bigger than the popular 35 mm types. A piece of greaseproof paper can be held in place over the opened back by means of an elastic band or sellotape. By adjusting the shutter to 'B' or 'T' and pressing the shutter control, the lens aperture can be kept open.

Experiment 11.7 (p. 28)
How does a camera work?
D.

Focusing camera.
Tissue screen.
Rubber band or sellotape.

Colour blindness

It is not necessary to go into a detailed discourse on colour vision theory at this level. From the careers point of view, however, it is imperative that at this early stage each pupil should be tested for colour blindness and the result notified to the careers master or form tutor. A pupil with serious colour blindness will not be able to become an aircraft pilot, railway engine driver, ships officer, electrician, dyer, etc. He should not, however, be made to feel that he is seriously handicapped.

The Ishihara Colour Vision Test cards should be available. The accompanying booklet gives the diagnosis of the degree of defect. The test should be administered to pupils individually. Pupils found to be deficient should be reassured by the fact that the defect is relatively common in males (about 1 in 12) and that there are still many careers opportunities where colour blindness is not important. Some pupils may not have the opportunity of studying genetics later on and they may be interested to know that the defect is handed on to males from their maternal parent (like haemophilia) although it is very rare in females and the female handing on the tendency will not herself be deficient.

Experiment 11.8 (p. 28)
Are you colour blind?
Ishihara Colour Vision Test Cards and Handbook.

Experiments 11.5, 11.6, 11.7, and 11.8 could be done by the pupils using the stations technique after the apparatus has been demonstrated by the teacher.

Vision

The experiments on vision which follow can all be done quickly with very little apparatus. Full instructions are given in the *Pupils' Textbook* (Experiments 11.9–11.16; pp. 28–31). Some additional notes are given below for several of the experiments.

Judgment of distance. Binocular vision is important to animals which prey on others. Animals which are hunted use mainly monocular vision. A homework exercise on this topic might ask the pupils to list animals of prey and typical victims, giving in each case a drawing of a frontal view of the animals' heads.

Range of vision (*field of view*) (Experiment 11.12, p. 29). It is imperative that the pupil focuses on a fixed point when range of vision is being plotted but this experiment should not be done in too precise a way or valuable time will be lost. Any pupil who suffers from migraine will be able to tell how the field of view is limited during an attack.

Persistence of vision (Experiment 11.14, p. 30). There are many opportunities here for homework topics.

Blind spot (Experiment 11.15, p. 30). An interesting variant of the experiment described in the *Pupils' Textbook* is to try to make a human head, two metres away, disappear. Keeping the left eye closed, focus the right eye on an object about seventy-five centimetres to the left of a pupil's face. The face will be found to disappear. Success requires a certain amount of patience. It is said that this method of 'beheading' gave Charles II a great deal of amusement!

Boxers, and cricketers batting at the wicket, soon realize the existence of the blind spot and take steps to avoid missiles from this direction.

Literature on the eye and vision

So you know about your eyes, Family Doctor Booklet. B.M.A., 47–51, Chalton Street, London, N.W.1.

Eye and Brain. The Psychology of Seeing, by R. L. Gregory. World University Library. Paperback.

Optical Illusions, by S. Tolansky. Pergamon Press.

Optical Illusions and the Visual Arts. Studio Vista.

Film

You and your eyes, featuring Jiminy Crickett. Walt Disney. Running time 10 min.

The ear and sound

It is helpful to have large wall charts of the regions of the human ear. These can be made by the teacher himself who can thus avoid the unnecessarily complicated anatomical detail which is usually given in commercial charts. Commercial models of the ear are generally rather small and expensive. Some teachers have made an art of making anatomical models on a large scale out of wire netting, and wire bent to shape, and then coated with papier mache. When painted, these three-dimensional models have found ready acceptance by pupils and they cost practically nothing. They have the merit that they can be made to look much less complicated than the commercial models at present available and which are really intended for use in connection with the much later stages of a science course.

The important regions of the ear are (i) the outer ear, (ii) the middle ear, and (iii) the inner ear.

The outer ear. Pupils should be asked about the part played by the sense of hearing in the animal world. Behaviour shown by dogs, elephants, etc. can be cited; the 'pricking' of the ears and the turning of the head to locate the source of sound should be referred to.

The ear flap (pinna) serves to catch the sound waves and direct them on to the ear drum. The cupping of the hand to the ear, and the use in former days of an ear trumpet by the deaf, can be mentioned here.

The middle ear. It should be made clear to the pupils that the three bones act as a lever system and magnify the vibrations of the ear drum.

The inner ear. This resembles a snail's shell of some three and a half turns. Sound vibrations travel up one side and down the other; the nerves along the Organ of Corti are each sensitive to one particular frequency (Figure 11.3).

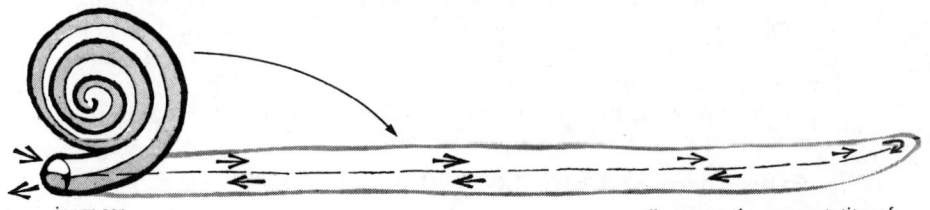

inner ear

diagrammatic representation of the movement of sound waves in the inner ear when it is uncoiled

FIGURE 11.3

Besides the items listed in this section, it could be expected that pupils will suggest many other ways of producing sounds: tapping with a ruler on tins, or bottles, or on the bench; shaking a tin of peas; vibrating a ruler; or twanging a rubber band. The class might be asked to list the various sounds they are accustomed to hear and to identify the origin and production of the sound. It will soon become clear that in each case the origin of the sound lies in some kind of vibration.

Materials that might be available in the laboratory for this part of the lesson include: a sonometer; organ pipe; recorder; door-bell chimes; tuning forks; signal generator and loud speaker.

Ask the pupils to bring in to school any stringed instrument they play. This should result in a collection of guitars and possibly a violin, which can all be used to do Experiment 11.17 (p. 32). If possible take the pupils to the music department to see the inside of a piano. The science department should possess a sonometer which merely consists of a wire stretched across a sounding box. It is possible to alter the length of the string which vibrates by moving a wooden wedge along the wire, and also to alter the tension.

Pupils should discover that the pitch is raised by (i) increasing the tension, (ii) decreasing the length of string which vibrates, and (iii) making the wire thinner or lighter.

To continue further the study of the production of sound, obtain the larynx of a sheep from the slaughter house. The vocal chords are easily seen.

Mention could be made of the 'breaking' of the voice of teenage males and the slackening of the vocal chords.

Experiment 11.18 (p. 33)

D.

1 sheep's 'pluck'.

Cathode ray oscilloscope and microphone.

In common with the effect of shortening the length of vibrating strings, shortening the length of a vibrating ruler results in a raising of pitch. Pupils should examine tuning forks and should note that forks which are shorter in length give higher pitched notes.

When a vibrating tuning fork is lowered to touch the surface of water in a basin, waves will be set up in the water.

Experiment 11.19 (p. 33)

Rulers.

Tuning forks.

A number of tuning forks could be put out and pupils could move from one to another comparing the notes they give and their characteristics. Tuning forks should be sounded by striking them against a resilient surface such as a rubber pad, or a rubber-heeled shoe, not against a hard surface such as that of the bench.

If the c.r.o. has been set up for the previous experiment it is interesting to sound a tuning fork in front of the microphone and to observe the trace it produces on the c.r.o. screen. A simple discussion of the difference between the effect produced by the tuning fork and that produced by the voice could follow. Even the most complex sound can be broken down into simple vibrations such as that of the tuning fork.

Ask pupils who play wind instruments, such as the recorder, clarinet, saxophone, cornet, etc. to bring them in and explain to the class how their instrument is played. They should discover that the longer the vibrating air column the lower the pitch of the note.

Experiment 11.20 (p. 34)
Wind instruments

P. 3 groups.

8 milk bottles.

8 test-tubes in racks.

8 gas jars.

Each group will take one of these and try to set up a musical scale.

Experiment 11.21 (p. 34)
The signal generator

D.

1 signal generator and loud speaker.

1 cathode ray oscilloscope.

As suggested in the film *Science for the Seventies* the following experiment can be done as an intra-disciplinary team teaching exercise. If possible, gather all the pupils in the second year together in, say, the school hall. As the pitch from the signal generator rises pupils are asked to stand as they lose the note. They are counted, and the frequency noted. This is repeated until the note is inaudible to all. The fact that there are still vibrations in the air is shown by the fact that there is still a trace on the c.r.o. The results are collected on a statistics board or can be represented on a histogram. The experiment should be repeated with all the pupils seated. The frequency of the note is gradually lowered, and pupils stand as it becomes audible to them. A normal distribution curve should be obtained in each case.

Experiment 11.22 (p. 35)
Your range of hearing

D.

Signal generator.

Cathode ray oscilloscope.

Loud speaker.

Statistics board (or other method of recording results).

In further discussion, the better pupils may be able to point out possible sources of error in this experiment. For instance, we are dependent upon the response of the loud speaker. It may well fail to respond to higher frequencies before our ears do. Of course, a good loud speaker as free as possible from resonances is required for this experiment.

Transmission of sound

The object of Experiment 11.23 is to show that sound waves require a medium in which to travel. An electric bell is sounded in a vacuum but the ringing is inaudible to the pupils.

For this experiment to work well it is imperative to have a good vacuum pump; it is also essential that the pump plate should be plane, that the bell jar is sealed to the plate with vaseline, and that thin wires are used to suspend and supply current to the bell. As air is removed from the jar the sound should become fainter and fainter but it should be obvious to the pupils that the gong is still being struck by the hammer.

Experiment 11.23 (p. 35)

D.

Vacuum pump.
Pump plate.
Bell jar.
Electric bell.
Laboratory power pack or other source of electrical supply.

The *Pupils' Textbook* mentions several examples of the transmission of sound through various media. Many other examples, besides those mentioned, may crop up in class: the Indian's ear to the ground listening for the pursuing cowboys' horses hoof beats; the train robbers listening for the sound of the approaching train transmitted along the rails; the silence necessary in the hunted submarine as sound travels so well in water. At this point some pupils may be interested to hear themselves as others hear them—on the tape recorder. We hear our own voices by the conduction of vibrations from our larynx through the facial bones to the middle ear. When we first hear our recorded voice we are astonished at the difference between what we think of as our true voice and the record. Reasons for the difference might be discussed. Reference could be made to the deaf Edison being able to hear his first recordings by picking up sound vibrations from the instrument through his teeth biting into it.

If any pupil in the class uses a hearing aid, opportunity might be taken of discussing how it works. Some of these instruments operate by bone conduction.

Taste, smell, and other senses

Experiment 11.24 and 11.25 (p. 37)
Can you detect different tastes?

P. 10 groups.

---

2 sets of beakers containing:
1. sucrose solution (sweet)
2. sodium chloride solution (salt)
3. citric acid solution (sour)
4. quinine sulphate solution (bitter)

Glass rods (4 per pupil).

Sweet and salty substances are tasted at the front of the tongue, acids at the side, and bitter substances at the back. There will, however, be variations. Some pupils will be able to detect all tastes at the front of the tongue, although acid and bitter tastes will be felt more strongly at the side and back respectively.

A point which might form the basis of discussion with the class is that as food enters the mouth it comes into contact firstly with the front of the tongue and it is this region which is most sensitive to taste.

Experiment 11.26 (p. 37)
Taste and smell

P. 10 groups.

Cubes of apple, onion, potato, and turnip.
Clip for the nose.

The above experiment shows how limited our sense of taste is and the role played by our sense of smell in detecting the flavour of food. The pupils will have had the experience of 'tasteless' food when they have had a heavy cold. Reference can be made to the effect of smoking on the taste buds of the tongue. A heavy smoker may have great difficulty in distinguishing between tastes and may even be unable to detect them at all.

Experiment 11.27 (p. 37)
Touch

P. 10 groups.

10 pairs of dividers *or* mounted needles.

Experiment 11.28 (p. 37)
Temperature and touch

P. 10 groups.

10 mounted needles.
Beakers of boiling water.
Beakers of ice.

Nerves

The skin contains numerous sensory receptors. The above experiments show:

1. that there are different nerve endings for the sensations of touch, heat, and cold,
2. that each kind of nerve ending is stimulated only by one sensation, and
3. that the nerve endings are not spread evenly over the skin. The points of the fingers and parts of the palm contain more nerve endings than the back of the hand.

Reflex actions   Reflex actions are generally inborn reactions. Experiment 11.29 demonstrates this type of reaction. They are immediate, automatic reactions and because of them the body is able to respond more quickly to changes in the environment than if the information had to be dealt with by the brain. They therefore play an important part in protecting the body from injury.

While most reflex actions are inborn, some can be developed by constant repetition. These are termed 'conditioned' reflexes. They can be demonstrated with fish in an aquarium. When fish are fed they rise to the surface of the water. To demonstrate conditioning, tap the side of the tank each time the fish are fed. They soon rise to the surface each time the side is tapped, even if no food is offered.

Reaction times

Experiment 11.30 (p. 39)

P. 10 groups.

10 stop watches.

The brain and spinal cord

P.T. Fig. 11.31 can be used to show the structure of the brain. Some of the functions of the main regions are shown on the diagram. There should be no attempt to learn the names of the regions of the brain.

The important points to note are:

1. the brain is a very complex organ,
2. it controls and co-ordinates the activities of the body,
3. it receives impulses from sense organs and sends out motor impulses as a result, and
4. it retains information.

Unit 12: The Earth and What We Get From It

Specific objectives

The objectives of this unit are that the pupils should acquire:

1. knowledge of some facts about the origin and structure of the earth,
2. knowledge of some facts about naturally occurring elements and ores,
3. knowledge of the reasons for the presence of these elements and ores in the earth,
4. further knowledge of the idea of order of activity in the elements,
5. knowledge of some facts about silica and silicates,
6. knowledge of possible means of forming metamorphic rocks,
7. some information about colours in minerals and glazes,
8. some information about the fossil fuels (coal, oil, and natural gas),
9. some information about the salts of the sea,
10. knowledge of some facts about the soil,
11. knowledge of some facts about micro-organisms,
12. the ability to form hypotheses from experimental observations using data derived from experiments on oxides, sulphides, and carbonates,
13. the ability to retrieve information about earth, fossil fuel, rock types, etc.,
14. the ability to use acquired knowledge and skills in solving a problem of identification of an unknown substance (malachite). (This involves both analysis of material to obtain information and a synthesis of the findings to provide a reasonable solution.),
15. further ability to use a key in identifying unknown creatures,
16. awareness of the importance of certain properties of minerals in the earth which allow them to be used as building materials,
17. interest in the need for conservation of fuel resources,
18. awareness of the importance of the sea as a source of minerals,
19. awareness of the place of micro-organisms, both useful and harmful, in the life of man,
20. various chemical and biological skills, and
21. some simple micro-biological techniques.

This is an important unit. The subject matter covered has seldom been given due attention in traditional syllabuses. There is a considerable amount of material in the unit and every effort must be made by the teacher to maintain the interest of the pupils by making the work as practical as possible. The study of rocks, even in the limited form suggested here, can be of absorbing interest if properly presented and it is hoped that a number of pupils will make this one of their hobbies as a result of the work of this unit. If at all possible, some field study should be undertaken. Geological expeditions can be arranged in association with the geography department. A collection of minerals should be made using, in the main, samples brought in by the pupils. Each specimen should be labelled with its name, composition, type, and location, and it adds interest if the name of the donor is also given.

It is suggested that, apart from field work, the *time allocation* for this unit should be thirty periods.

Order of development

The unit is developed in the following order in the *Pupils' Textbook*:

1. First the origin of the earth is discussed briefly. There are, of course, conflicting theories, and it is made clear that some of the evidence we have comes from a study of rocks.
2. The various types of rocks recognized by geologists are considered —igneous, sedimentary, and metamorphic—and their origin and age are discussed.
3. The structure of the earth is briefly dealt with.
4. The nature of minerals from which metals are obtained is considered and the following questions are asked. Why do comparatively few elements occur as such in nature? Why are many metallic minerals sulphides, oxides, or carbonates? Why are most of the ancient rocks silicates and why cannot they be used to obtain metals? These points are developed on the basis of the activity series. It is realized, of course, that with pupils at this stage of knowledge and maturity only partial answers to these important geochemical questions can be given. More precise solutions involve thermodynamic considerations which are beyond the comprehension of the pupils but it is hoped that the simple treatment given will contain sufficient experimental observation to provide the basis of a reasonable hypothesis.
5. The general principles of the smelting of ores are considered without going into any detail. Here again it has been necessary to adopt a very much simplified approach which, however,

should be found useful as a basis for more detailed study if the pupils go on to more advanced work.

6. The pupils are given an unknown mineral, malachite, to investigate for themselves.
7. The importance of silicates for building materials, pottery, and glass is considered.
8. The origin and methods of obtaining the fossil fuels—coal, oil, and natural gas—are dealt with.
9. Minerals obtainable from the sea are investigated.
10. The formation of soil and the different types of soil are studied.
11. The soil is the home of myriads of creatures. The presence and value of bacteria in the soil are considered.

Looking at rocks

The object of the examination of rocks by the pupils is to lead them to the hypothesis that the earth was at one time in a molten state. They examine rocks which are obviously crystalline, such as quartz, granite, and calcite (others can be supplied from the school collection if desired), and the question is asked 'What evidence about the origin of these rocks can be obtained from their crystalline nature?' The pupils should know from their work in Unit 5 that crystals are formed either from a melt or from a saturated solution. (It is also possible for crystals to be formed direct from the vapour state but this will probably not be known to the pupils.) The fact that there are volcanoes which are still active indicates that the inside of the earth is still molten.

Experiment 12.1 (p. 41)
Looking at rocks

P. 10 groups.

Several specimens of quartz, calcite, granite, or other crystalline minerals. (They can be circulated round the class.)
Knives.
Hammers.
Metal sheets on which the crushing of the mineral can be tested.
Test-tubes.

Sediment

The idea of sedimentation is illustrated by taking a mixture of fine gravel, sand, and powdered clay, shaking this with water in a gas jar, and allowing it to settle, as outlined in Experiment 12.2 (p. 43). The gravel should sink to the bottom, the sand should form a layer on top of the gravel, and the powdered clay should remain in suspension for some time. If salt water is used instead of tap water the clay deposits more quickly. This is due to the fact that the clay particles (or the finer of them) are of colloidal dimensions and are charged. The ions in the salt solution neutralize the charge and bring about precipitation. The pupils may have met this idea in Unit 5, but if not they should merely make the observation. The pupils are asked to look at sands from different sources. Sand is a product of degradation of silicate rocks. It always contains impurities. Pure sand is silicon dioxide and should be perfectly white (or colourless). The nearest approximation to pure silicon dioxide is silver sand which is found on some of our beaches and on the beds of some Highland streams. Commonly, sand is yellow to brown in colour, due to the presence of varying quantities of iron compounds. In some areas, sand which has originated from volcanic rocks is black.

Experiment 12.2 (p. 43)

P. 10 groups.

20 gas jars.
Mixture of fine gravel, sand, and powdered clay.
Sodium chloride solution (about 10 g per litre).
Hand lenses (may be shared).
Samples of different sands.

The average density of the earth's crust is about 2.7 g cm^{-3}. The pupils will be interested in finding the densities of various rocks. Most of these will fall between 2 and 3 g cm^{-3} but igneous rocks are generally more dense than metamorphic and metamorphic than sedimentary. The pupils should find the mass and volume of a small specimen.

Experiment 12.3 (p. 44)

P. 10 groups.

Butchart balances.
10 measuring cylinders (100 cm^3).

Specimens of rocks such as quartz, granite, slate, basalt, marble, limestone. (The blocks of slate included in the Nuffield Materials Kit can be used; their volume may be calculated by measurement of length, breadth, and depth.)

If you are likely to be pressed for time this experiment could be omitted, or carried out only by the faster-working pupils.

Elements in nature

The pupils have so far been considering the occurrence of elements in nature and will have come to the conclusion that many of them do not occur uncombined. Metals with electrode potentials greater than hydrogen (i.e. lower than hydrogen in the activity series) are rarely found in the free state. Pupils should be guided to this conclusion by considering the readiness with which metals combine with oxygen and sulphur, two non-metallic elements which occur in the earth's atmosphere and in its crust, respectively.

The following table, which may be useful to the teacher, gives the approximate percentage by weight of the sixteen most abundant elements, *free* and *combined*, in the earth's crust. It is not advisable to give this table to the pupils as they tend to confuse the terms *free* and *combined*. Thus, some of the 49.5 per cent of oxygen is free and some is combined but there is no free silicon or aluminium.

Abundance of elements in the earth's crust			
Element	% by weight	Element	% by weight
Oxygen	49	Hydrogen	
Silicon	26	Titanium	
Aluminium	8	Chlorine	
Iron	5	Phosphorus	Each less than 1
Calcium	3	Manganese	
Sodium	3	Carbon	
Potassium	3	Sulphur	
Magnesium	2	Barium	

As oxygen and sulphur are adjacent elements in the same group in the Periodic Table it would be expected that they would react in a similar way with metals. Many metallic elements combine directly with sulphur when heated, as they do with oxygen.

In the experiments on the direct combination of metals with sulphur a mixture of the metal with powdered sulphur (or flowers of sulphur) is made and heated in a small test-tube. There is evidence of combination because of the considerable amount of energy given out and the appearance of the product. With silver a black stain is produced where the foil is rubbed with sulphur. The sulphur is best rubbed on with a cork. Similarly, a small drop of mercury can be rubbed with sulphur in a mortar.

Warning: Mercury vapour is poisonous. In order to avoid the presence of much mercury vapour in the laboratory, this experiment could be demonstrated in a fume chamber. After the experiment the mercury and its sulphide should be dissolved in concentrated nitric acid, the liquid diluted, and poured down the sink. Mercury itself should never be poured down the sink.

Experiment 12.4 (p. 44)
The action of sulphur on metals

P. 10 groups.

Powdered sulphur or flowers of sulphur.
Iron filings, zinc foil or filings, copper powder or filings, lead shot, aluminium powder or filings, magnesium ribbon (small pieces).
20 small test-tubes (7.5 cm × 1 cm).
Bunsen burners.
Silver foil.
Mercury ⎱
Mortar ⎰ for demonstration.

It is not necessary for all groups to try all the metals; two per group would be sufficient. Results for the class can be pooled.

The simplest method of examining the combination of metals with oxygen is to use the apparatus shown below (Figure 12.1).

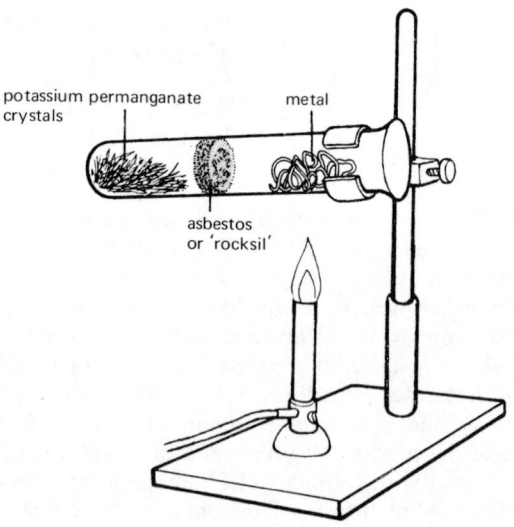

potassium permanganate crystals

metal

asbestos or 'rocksil'

FIGURE 12.1

The oxygen is obtained by heating the potassium permanganate crystals at the end of the tube. The oxygen passes over the heated metal. Some idea of the vigour of the reaction can be obtained from the effect on the tube. Do not use magnesium powder in this experiment.

Experiment 12.5 (p. 44)
The action of oxygen on metals

P. 10 groups.

Potassium permanganate (solid), magnesium ribbon, calcium (freshly scraped), aluminium foil, zinc foil, lead foil, iron filings, copper filings, silver foil, asbestos, or 'rocksil'.
20 test-tubes.
10 retort stands.
Bunsen burners

It is not necessary for each group to examine more than two metals.

Carbonates

Carbonates are sedimentary rocks. Carbonates other than those of sodium and potassium give off carbon dioxide when heated and leave the metal oxide. In the case of carbonates of metals of which the oxide is unstable (e.g. silver), both oxygen and carbon dioxide are given off and the metal is left. We are not, however, interested in this case here.

In Experiment 12.6, the pupils are not told which gas is given off; they have to discover this for themselves by applying appropriate tests. They will probably see first whether the gas burns or supports combustion. The fact that they are heating carbonates should suggest to them that it would be worth while testing for carbon dioxide.

When calcium carbonate in any of its forms is strongly heated, carbon dioxide is evolved and calcium oxide is left. This oxide combines vigorously with water to form calcium hydroxide and, if the oxide is not swamped with water, sufficient heat is evolved to boil the water. The common name for calcium oxide is *quicklime* and for calcium hydroxide, *slaked lime*. These names are interesting and expressive, but it is preferable that the pupils remember the chemical names rather than the common ones.

Zinc oxide is yellow when hot and white when cold.

Pupils should be able to deduce for themselves that there is some connection between the position of a metal in the activity series and the ease of decomposition of its carbonate.

Experiment 12.6 (p. 45)
Heating carbonates

P. 10 groups.

30 test-tubes (7.5 cm × 1 cm).
10 pairs crucible tongs.
10 watch glasses.
Calcium carbonate (powdered chalk), zinc carbonate, lead carbonate, marble chips.
Lime water.
Bunsen burners.

Experiment 12.7 (p. 46)
More experiments with carbonates

P. 10 groups.

20 test-tubes.
Lime water.
Sodium carbonate, calcium carbonate, magnesium carbonate, zinc carbonate, barium carbonate, potassium carbonate.
Dilute hydrochloric acid.
Dilute acetic acid.

How metals are obtained from their ores

The next section deals briefly with the extraction of metals. In general the process resolves itself into the reduction of oxides usually with carbon or carbon monoxide, although electrolytic reduction has to be used in the case of some metals, the oxides of which cannot be reduced by other means. The pupils show first that sulphide ores (typified by iron pyrites) are normally converted into oxides when heated in air. The following experiment shows how the oxide produced can be reduced with carbon. The heating of powdered iron pyrites is best carried out on a strip of asbestos paper.

Experiment 12.8 (p. 46)
Heating iron pyrites in air

P. 10 groups.

Asbestos paper.
Bunsen burners.
Powdered iron pyrites.

Experiment 12.9 (p. 46)
Heating oxides with carbon

P. 10 groups.

Asbestos paper.
Powdered charcoal.
Lead oxide, iron(III) oxide, copper(II) oxide.
Dilute nitric acid.
10 watch glasses.

Malachite is copper(II) carbonate. With the experiments they have recently performed fresh in their minds the pupils should be able to find this out for themselves without any guidance from the teacher. The pupils should be shown some of the mineral but should be given commercial powdered copper(II) carbonate to experiment with. They can be told that it is the powdered mineral, although actually it is a basic carbonate. They should be guided towards the following procedures: heat the substance, show that carbon dioxide is given off, and note the colour change. They might then confirm that it is a carbonate by acting on it with dilute hydrochloric acid; they should take some of the black powder obtained by heating the carbonate and reduce it by heating with charcoal, testing the copper coloured residue for copper by means of dilute nitric acid. They might also observe the green flame colouration which can be obtained at almost any stage of the proceedings.

Experiment 12.10 (p. 48)
An investigation of malachite

P. 10 groups.

Malachite for demonstration.
Copper(II) carbonate.
10 test-tubes, 7.5 cm × 1 cm.
20 test-tubes, 15 cm × 1.5 cm.
Dilute hydrochloric acid.
Lime water.
Powdered charcoal.
Asbestos paper.
Bunsen burners.

Silica and silicates

It is not realistic to go very deeply into the chemistry of silicates at this level. We want the pupils, however, to realize that they are comparatively unreactive, being very resistant to the action of water and dilute acids. This explains why the world's highest mountains are composed of silicates and why silicates form the basis of building materials. This chemical unreactivity is due mainly to the structure of the silicates. For a simple treatment of the subject, suitable for Certificate pupils at this stage, the teacher is recommended to see *Chemistry Takes Shape*, Book II, by Johnstone and Morrison (Heinemann Educational Books Ltd).

Experiment 12.11 (p. 48)
Some experiments with silicates

P. 10 groups.

Specimens of sand, clay, mica, and felspar.
10 test-tubes.
Dilute hydrochloric acid.

Experiment 12.12 (p. 49)
Some experiments with clay

(This experiment should be carried out in collaboration with the art department.)

P. 10 groups.

Potter's clay.
Kiln (a good muffle furnace will answer the purpose).

Experiment 12.13 (p. 49)
Glazes

P. 10 groups.

Small tiles made in Experiment 12.12.
Sodium chloride solution (about 100 g sodium chloride in 500 cm³ water).
Red lead.
Powdered flint (or silver sand, or precipitated silica).
Cobalt(II) nitrate.
Other substances which can be used in place of cobalt(II) nitrate: manganese(II) sulphate, iron(III) chloride (or sulphate), chromium(III) sulphate, nickel(II) sulphate.

Coal

We now pass to a consideration of the fossil fuels and, of these, the most recent is peat. The school should have available specimens of different types of coal as stated in the *Pupils' Textbook*.

Experiment 12.14 (p. 50)
Peat

P. 10 groups.

Small pieces of peat. This should be, if possible, the type used for burning by crofters and should show some evidence of plant roots, etc.

Experiment 12.15 (p. 50)
Heating coal

P. 10 groups.

Small tin with hole in lid.
Pieces of coal.
Bunsen burner.

Pupils should observe that an inflammable gas is given off. They should examine the residue and compare it with the original coal.

Experiment 12.16 (p. 50)
Making coal gas

P. 10 groups.

10 sets of apparatus as shown in P.T. Fig. 12.13.
Bunsen burners.
Indicator paper.

Experiment 12.17 (p. 51)
Heating wood

P. 10 groups.

10 sets apparatus as for Experiment 12.16.
Bunsen burners.
Indicator paper.

Experiment 12.18 (p. 52)
Distilling crude oil

P. 5 groups.

5 sets of apparatus as shown in P.T. Fig. 12.15.
5 evaporating dishes.

The sea

The object of the following experiment is to show that sea water contains more dissolved solid than distilled water or tap water and then to show that the salt of the sea is mainly sodium chloride. This will introduce the pupils to the flame test for sodium and the silver nitrate test for chloride ion.

The experiment should be carried out with small quantities of the different waters in order to save time. The evaporation is best done on glass rather than in porcelain basins so that the residue shows up. A black glass tile is useful with evaporation being done by an infra-red lamp.

Experiment 12.19 (p. 52)
Evaporating water

P. 10 groups.

Samples of distilled water and sea water. (De-ionized water cannot be used in place of distilled water. If the school is far from the sea, and sea water is not readily obtainable, tablets of sea salt can be purchased from chemical suppliers and dissolved in water according to the supplier's instructions; alternatively a dilute solution of sodium chloride—2.5 g of sodium chloride in 100 g water—can be used.)
Black glass tiles, or watch glasses, or microscope slides.
Infra-red lamp.
10 test-tubes.
Pieces of clean iron wire, nichrome wire, or pencil lead for flame test.
Silver nitrate solution.
Ammonia solution.

The soil environment

The soil is the habitat of millions of organisms which require food, oxygen, and protection. The object of the present study is to see how the soil provides these requirements.

A brief account of the formation of soil by weathering and the subsequent colonization of the mineral fragments by plants should show how soil particles are formed, why they differ in size, and why soils in different regions vary in composition. Some of the chemical and physical properties of soil can be investigated with the purpose of (a) finding out why the soil is a suitable environment for so many organisms, (b) linking the numbers of organisms present in the soil with its physical conditions, and (c) comparing the numbers and kinds of organisms in a variety of soils.

Collecting soil samples

Three methods of collecting soil samples are given in the *Pupils' Textbook*—two on p. 55 and the third on p. 56. If there is any ground nearby, the pupils should be taken out and shown how to collect the soil by each method. If an auger is not available, a long column of soil can be obtained by pushing a piece of metal pipe into the earth and withdrawing it. The soil can be extracted by pushing a stick down the pipe.

Soil should be collected by pupils from a variety of sites. They should be reminded to label each sample and note any relevant information about the area from which they collected it, e.g. if there were many plants or few in the surrounding region, if the soil came from a cultivated or a natural area, or if it shows any signs of having been flattened by animals walking over it.

To get a distinct variation in soil types, soil should be collected from a woodland area (rich in humus), an area near the coast (sandy soil), a hillside or near to a river (often clay), and a cultivated field (often loam). As the kind of soil varies from district to district a short preliminary investigation should be made to ensure that the soils to be used by the pupils show a variation in particle size and composition.

Experiment 12.20 (p. 53)

P. 10 groups.

Microscopes 10, if possible; otherwise reduce the number of groups, or share.
Bench lamps.
Microscope slides.
Cover slips.
10 dropping pipettes.
10 mounted needles.
Samples of soil from different sources.

Experiment 12.21 (p. 54)

P. 10 groups.

10 measuring cylinders.
Samples of soil.

Experiment 12.21 illustrates a method of separating soil according to the sizes of the particles present and is similar to the experiment on sedimentation (Experiment 12.2). The particles separate as shown in P.T. Fig. 12.5.

The pupils should be able to predict the comparative sizes of the particles in each layer and then go on to test their predictions.

Experiment 12.22 (p. 54)

P. 10 groups.

10 microscopes.

Microscope slides.
Cover slips.
Bench lamps.
Labelling pencils.
Measuring cylinder and separated soil from Experiment 12.21.

Water, humus, and particle size

For Experiments 12.23, 24, and 25 the class should be divided into groups, each group using soil collected from a different region. Each experiment is a stage in the analysis of the soil sample. As the experiments are done quantitatively care should be taken not to lose any soil between each experiment.

Experiment 12.23 (p. 55)

P. 10 groups.

10 large tin lids.
Samples of fresh soil. (Different groups should work with different samples.)
Incubator or oven.
Balance (a chemical balance will be necessary).
10 spatulas.
Labelling pencils.

Experiment 12.24 (p. 56)

P. 10 groups.

Tin lid plus sample of soil dried in Experiment 12.23.
10 bunsen burners.
10 tripod stands and gauzes.
Balance.

Experiment 12.25 (p. 56)

P. 10 groups.

Sample of soil used in Experiments 12.23 and 12.24.
10 pestles and mortars.
10 sieves.
Balance.

Experiment 12.26 (p. 56)

P. 10 groups.

Air-dried samples of top and sub-soil gathered from the same spot.
10 bunsen burners, tripod stands and gauzes.
Balance.
Leaf litter is formed in the top soil and so this layer will have more humus than the sub-soil.

Experiment 12.27 (p. 56)

P. 10 groups.

10 measuring cylinders (500 cm³).
Tin can containing sample of fresh soil.

The results can be calculated as follows:

Volume of water in cm³ $\qquad$ $= 200$
Volume of water + soil $\qquad$ $= 200 + x$
Volume of water + volume of can $= 200 + y$ cm³
Volume of air in sample $\qquad$ $= y - x$ cm³

Alternative method for comparing the amount of humus in soils

'20 volume' hydrogen peroxide is used in this experiment. As this liquid can cause blisters on the skin, the experiment should be carried out as a demonstration.

Allow samples of soils to air-dry in the laboratory for two or three days. The following procedure should be carried out for each kind of soil and the results compared.

Put 1 g of soil into each of two boiling tubes, A and B. Add 10 cm³ of '20 volume' hydrogen peroxide to A and add 10 cm³ of water to B. Warm both gently in a water bath. When the frothing in A appears to have ceased, remove the tubes from the bath and add about 20 cm³ of water to each tube. Allow the contents to settle. Compare the colour of the soil in A with that in B. The contents of A should be paler owing to the bleaching of the humus by the hydrogen peroxide. A slight difference in colour indicates a small amount of humus in the sample, a considerable difference shows that a large amount was present.

Animals in the soil

Having considered types of soil we pass on to look at the animals which are to be found in a soil environment. The importance of the constituents of the soil to the animals living in it should be discussed with the class as a whole and then summarized in a table. One possible form of the table is as follows:

Substance in the soil	Use
Mineral particles	To give protection.
Water	To keep the skins of many animals, e.g. earthworms, moist for gas exchange.
Humus	To provide food.
Air	To provide oxygen for respiration.

Experiments 12.28, 29 and 30 describe methods of extracting animals from the soil. Animals will move out of the soil when the conditions are altered. When a light source, e.g. a bench lamp, is placed above a sample of fresh soil or leaf litter in a Tullgren funnel, the organisms will move away from the surface layers, pass through the soil, and fall into the collecting jar. It is difficult to collect animals which live in soil water by this method. These can be collected using a Baermann funnel.

Experiment 12.28 (p. 59)

P. 4 groups.

Soil and leaf litter.
Newspaper and/or an enamel tray.
4 hand lenses.
4 small dishes, e.g. crystallizing dishes.

Experiment 12.29 (p. 60)

P. 4 groups.

Soil and leaf litter.
Formaldehyde or alcohol.
4 large plastic funnels with gauze to fit.
4 beakers.
4 bench lamps.
4 hand lenses.
4 microscopes.
Microscope slides and cover slips.
4 dropping pipettes.

If plastic funnels are not available a cone can be made from a large sheet of paper.

Do not cover the gauze completely with soil, and do not pack the soil too tightly into the funnel; a space should be left between the inside of the funnel and the soil to allow air to circulate.

Experiment 12.30 (p. 60)

Apparatus as for Experiment 12.28.

Experiment 12.31 (p. 61)

P. 4 groups.

4 filter funnels.
Muslin.
Rubber tube, about 5 cm long to attach to stem of funnel.
4 clips.
4 beakers.
4 bench lamps.
4 microscopes.
Microscope slides and cover slips.
Formaldehyde or alcohol.
Soil.

Although bacteria cannot be seen in the soil their presence can be detected by the production of carbon dioxide.

Experiment 12.32 (p. 61)

P. 4 groups.

4 sieves, 2 mm mesh or less.
4 crucibles.
4 bunsen burners, tripod stands and gauzes.
Muslin.
Thread.
8 conical flasks with bungs to fit.
Labelling pencil.
Soil.
Bicarbonate indicator (see *Science for the Seventies*, Teachers' Guide, Book 1, p. 161).

Micro-organisms

Sterile Petri dishes are required for Experiments 12.33 and 12.35. These should be prepared in advance and given out to the pupils during the course of the lesson. Ideally each pupil should have a Petri dish; however, one dish can be shared between two, or possibly four if numbers are very large. For Experiment 12.33 ten Petri dishes will be required for a class of twenty pupils if one dish is shared between two pupils. Half the class can use fresh soil, the other half roasted soil. For Experiment 12.35 divide the class into groups of four. Each group can treat a different dish. In this case five Petri dishes are required for each class.

Disposable sterile Petri dishes can be obtained. They can only be used once for bacteriological work.

Glass Petri dishes can also be used. They can be sterilized after use and used again for bacteriological work.

Agars can be obtained from Oxoid Division, Oxo Ltd., Southwark Bridge Road, London, S.E.1.

Preparation of a Petri dish

Two agar tablets are required for each Petri dish. Add 10 cm³ of distilled water and two agar tablets (e.g. blood agar base) to a series of test-tubes. Allow the tablets to soak for 10 minutes. Plug each tube with cotton wool and sterilize the agar by heating the tubes in an autoclave or pressure cooker at 15 lb per square inch for 20 minutes. At the same time sterilize a flask of distilled water in the pressure cooker. This will be used in Experiments 12.33 and 12.35. If a pressure cooker is used allow the temperature to drop gradually to about 70° C and then remove the tubes. If the pressure is reduced too quickly, the contents of the tubes spurt out and about half the agar can be lost. Pour the contents of each tube into a sterile Petri dish. To do this lift the lid at one side to a height of 2 cm. If the lid is completely removed bacteria from the air might settle on to the base of the dish and so contaminate the agar. Lower the lid. Carefully rotate the dish so that the agar spreads evenly over the base. Allow the agar to set.

Experiment 12.33 (p. 61)

P. For numbers of groups see above.

Sterile Petri dish and agar.
Labelling pencil.
Incubator.
Crucible.
Bunsen burner, tripod stand, and gauze.
Water, sterilized then cooled.
Soil.
Roasted soil.

Experiment 12.34 (p. 62)
Can soil bacteria grow in any other food?

P. 10 groups.

20 conical flasks, 100 cm³.
Cotton wool.
Labelling pencil.
Fresh soil.
Roasted soil.
Milk.

Experiment 12.35 (p. 62)
Do other substances contain bacteria?

P. Number of groups, see above.

Sterile Petri dishes and agar.
Labelling pencil.
Incubator.
Various liquids: tap water, milk, pond water, sterile water +
nail scrapings.

Handling of contaminated plates

Contaminated plates, especially those containing soil micro-organisms, may contain pathogenic bacteria and so must be handled with great care. Bacteria can be killed, but preserved, by placing filter paper soaked in formaldehyde in the contaminated Petri dishes one or two days before the dishes are handled. Care should still be taken by the pupils not to remove any lids from Petri dishes. As an additional precaution the lids can be anchored to the bases with sticky tape.

Emphasis should be placed on washing hands after handling contaminated dishes. All hand-to-mouth operations, such as licking labels, should be avoided. The lids and bases of contaminated dishes should be placed in a bucket of 10 per cent lysol after use. The agar floats out of the dishes and can thus be disposed of. Disposable Petri dishes can be placed in an incinerator or can be destroyed by autoclaving.

Other micro-organisms

Fungal spores can be readily obtained from the dust in the laboratory. A mycelium will grow on damp bread in a few days. Different fungi will colonize the substrate and so it should be examined over a period of two to three weeks. The fungi which colonize the bread secrete enzymes which digest the starch. The digested food is absorbed into the hyphae. As a result, the amount of bread will decrease. The detailed structure of fungi need not be considered at this stage of the course. The 'animal-like' method of feeding and the lack of the characteristic 'plant-like' features can, however, be discussed.

Experiment 12.36 (p. 62)

P. 10 groups.

10 plastic boxes or Petri dishes.
Microscopes.
Bench lamps.
Microscope slides and cover slips.

10 mounted needles.
10 droppers.
Bread.

Experiment 12.37 (p. 62)

P. 10 groups.

10 plastic boxes or Petri dishes.
Incubator.
10 measuring cylinders, 100 cm³.
White bread, brown bread, pieces of fruit (orange or tomato), cheese.

The optimum conditions for the growth of fungi can be found by exposing different kinds of food, e.g. white bread, brown bread, and pieces of fruit to the air. The food which proves to be the most successful substrate (this is often brown bread) can then be used to find the optimum temperature and/or the optimum volume of water.

The latter can be carried out as follows:
Label four Petri dishes, A, B, C, and D.
Add the same amount of brown bread to each.
Add no water to A.
Add 10 cm³ of water to B.
Add 20 cm³ of water to C.
Add 50 cm³ of water to D.
Add some dust to each, replace the lids, and leave the dishes in a dark cupboard for about a week. Compare the amount of fungi growing on each. The optimum temperature for growth can be found by leaving dishes containing contaminated bread in a refrigerator, in a dark cupboard, in an incubator set at 32° C and in an oven set at 80° C.

Experiment 12.38 (p. 62)

P. 10 groups.

10 flower pots or yoghurt containers.
Soil.
Microscopes and bench lamps.
Microscope slides and cover glasses.
10 dropping pipettes.
Cress seeds.

The fungus *Pythium* grows on seedlings, such as cress seedlings, which have been planted close together and have been over-watered. Fill a flower pot or yoghurt container with soil and cover the surface thickly with cress seeds. When the seedlings grow, water them every day. The fungus produces a mycelium inside the stem of the seedling, which eventually withers and dies.

Useful micro-organisms

Micro-organisms are thought by many to be harmful to man. Many are, however, useful, and the remainder of this unit is concerned with a few of these micro-organisms.

Living yeast can be obtained from a brewery or a bakery. A culture medium for yeast can be made by adding 2 g of agar to 100 cm³ of a 2 per cent malt extract solution. The liquid should be sterilized and then inoculated with yeast.

Experiment 12.39 (p. 63)

P. 10 groups.

Microscopes.
Bench lamps.
Microscope slides and cover slips.
10 droppers.
Yeast.

Experiment 12.40 (p. 63)

P. 10 groups.

Plain flour.
Yeast.
Sugar.
Mixing bowl.
Oven or incubator.

A simple recipe for bread dough is as follows:

80 g plain flour
1 g dried yeast
2 g sugar
50 cm³ warm water.

Mix the dry ingredients and then add the water. Mix a second sample without yeast. Place each lump of dough into a bowl, cover, and after some minutes compare. The help of the homecraft department might be sought to allow the pupils to bake their products.

Experiment 12.41 (p. 63)

P. 10 groups.

10 test-tubes.
Balloons.

10 water baths.
Labelling pencil.
10 per cent glucose solution.
Yeast.
Bicarbonate indicator.

The glucose and yeast solution may be made by adding 4 g of dried yeast (or double the amount of living yeast) to 100 cm³ of 10 per cent glucose solution.

When yeast and sugar are mixed together carbon dioxide is liberated. The importance of the controls should be discussed. If the flask containing yeast and sugar solution is left for a few days alcohol can be detected in the flask. This can be separated from the liquid by distillation.

Experiment 12.42 (p. 63)

P. 5 groups.

Discs impregnated with *Bacillus subtilis*.
Blood agar base.
Nutrient broth No. 2.
Penicillin disc (5 units).
5 Petri dishes (one per group).
Labelling pencil.

Plate each disc with sterile blood agar base. Add a disc impregnated with *Bacillus subtilis* to 10 cm³ of Nutrient broth No. 2 in a sterile test-tube or a sterile McCartney bottle. Plug the tube, or replace the cap immediately to prevent contamination. This volume is enough for two Petri dishes. Incubate the culture at 37° C for 24 hours.

Shake the contents of the tube, and using a sterile pipette transfer about 5 cm³ to the surface of the blood agar in the prepared Petri dish. Rotate the dish so that the whole surface is covered. Dry the plates by placing them, with the lids off, in an inverted position in an incubator at 37° C. Discs impregnated with *Penicillin* can now be placed on the surface of the agar. The bacteria appear as a whitish covering over the surface of the agar. However, clear zones, or 'halos' appear round the Penicillin discs.

Unit 13: Support and Movement

Specific objectives

The objectives of this unit are that the pupils should acquire:

1. knowledge of what a force does,
2. knowledge that change of motion only comes about because of unbalanced forces,
3. knowledge that friction is always a resisting force,
4. knowledge of certain facts about gravity,
5. knowledge that the newton (N) is a unit of force and can be measured by a spring balance,
6. knowledge that the lever is a 'force multiplier',
7. knowledge that forces occur in pairs,
8. knowledge of the joule as a unit of work: 1 joule (J)=1 newton metre (N m),
9. knowledge of the ideas of motion energy and stored energy,
10. knowledge that a machine is an energy transformer but not an energy multiplier,
11. knowledge of some facts about support in plants and animals,
12. knowledge of some facts about muscular effort and the forearm as a lever,
13. ability to build the concept of force from a set of related facts,
14. ability to formulate the 'law of the lever' from a set of observations,
15. ability to develop a theory to explain observed phenomena (stability and leg arrangements in animals),
16. ability to apply the above knowledge to a new problem situation,
17. awareness of the need to postulate ideal conditions in order to formulate satisfactory physical concepts (e.g. movement without friction and ideal machines),
18. awareness that in the absence of external forces, uniform motion in a straight line is as probable as a state of rest,
19. awareness of the anomalous posture of man in relation to his structure, and
20. awareness of the fact that any machine must waste some of the energy input.

It is suggested that the *time allocation* for this unit should be twenty-four periods.

Order of development

In this unit we attempt to establish the concepts of force and work in an operational manner by relating them to various situations including the human frame. We then go on to study the suitability of various plants and animals to support the forces that they experience.

The order of development in the *Pupils' Textbook* is as follows:

1. The idea of force.
2. Work and energy.
3. Support in plants.
4. Support in animals.
5. Muscles and their action.

The idea of force

Until the pupils are given the opportunity to find out about forces in an organized way their ideas are open to many misconceptions. They will readily understand that forces as such cannot be seen; we are made aware of them only by their effects, as with so many other concepts in science. The sense of touch, coupled with discomfort in the muscles tells us when we are applying a force. Apart from these tangible forces there are others (less obvious) which pupils do not readily recognize and which are more liable to be taken for granted, such as, for example, the force of gravity. Because nothing seems to happen when a stone is lying on the ground the pupils are apt to think that no forces are acting, whereas in fact the earth is pulling the stone and the stone is pulling the earth. On the other hand, because in the case of an ice-hockey puck moving over the surface of an ice rink there is obviously something happening, pupils are tempted to think that a force is being continuously applied. This very common misconception arises from the fact that, in their experience, to maintain an object such as a garden roller or a car in motion a continuous force is required. This, of course, is due to the large force of friction which has to be overcome.

We set out in this unit to give the pupils experience of a number of different types of force and to correct wrong impressions, thus forming a good basic foundation for further work in mechanics.

The non-specialist physics teacher should bear in mind the following fundamental facts. A stationary object is at rest because forces acting on it are exactly balanced. They have no resultant effect. This does not mean that no forces are acting—only that they are balanced. Thus a stone lying on the ground is acted on by the force of gravity and also by an exactly equal and opposite force exerted by the earth on the stone. In the case of a body moving in a straight line with steady motion again there is no unbalanced force acting on the body. With the ice-hockey puck moving over the ice there is no force attracting it or pushing it, and the force of friction which opposes motion and slows the puck down is negligible. Thus a body which is at rest, or one which is moving in a straight line with steady motion (i.e. unvarying speed) may be acted upon by balanced forces, or no force at all. It cannot be acted upon by an unbalanced force as this would cause it to move, or to change its motion (i.e. increase or decrease its velocity, or cause its path to change in direction, which really amounts to the same thing). Thus if a stone is pushed over the edge of a cliff, then the unbalanced force of gravity would cause the stone to accelerate towards the beach below. If the ice-hockey puck were propelled over a concrete pavement instead of over ice it would soon slow down and come to a halt because of the unbalanced force of friction (unless it was rolling when friction does not apply in the same way).

The effect of forces on the shapes of bodies

Keeping in mind the model of the structure of matter in the solid state which was developed and used in Book 1, pupils are given a number of materials which they can push, compress, pull, stretch, twist, and deform. It is not intended that a detailed discussion on bonding should take place here but the effects of forces on solids should be interpreted in terms of the nature of the molecules and the forces between them.

Pupils should consider the spacing of the molecules and the type of molecule in the case of substances like rubber and plasticine. What happens to the molecules in an iron or copper wire when a heavy weight is hung on to it?

Questions to frame in connection with these simple experiments could also include:

In which material tested do the molecules exert the strongest forces on each other, judging from your efforts to pull them apart?

Do metal molecules exert strong forces against you when you try to push them closer together?

Experiments 13.1 – 13.3 (pp. 66–7)

P. 10 groups.

Plasticine.
A compression spring (such as a bed spring).
An extension spring.

Elastic band.
Foam rubber block (or plastic sponge).
Copper wire.
Steel wire.
Rigid support for wires.
Large masses (1 to 20 kg).

It is not necessary for these materials to be supplied for each group;
pupils can move about the laboratory.

The force of gravity

When pupils are asked why an object falls to the ground some usually
suggest that it is due to the weight of air above the body; others connect
it with the earth's magnetism. Many pupils also think that a heavier
body will fall faster than a light one. The purpose of this and the
following experiments is to clear away these false impressions.

The rate at which bodies fall can be examined by rolling a golf ball
and a marble off the edge of the bench together. A ruler can be used
to ensure that both roll over the edge at the same time. Only one sound
should be heard as both objects reach the floor together.

Experiment 13.4 (p. 67)

P. 10 groups.

10 golf balls.
10 marbles.
Rulers.

A similar experiment may be demonstrated by allowing a marble
and a small ball-bearing to drop down a long tube. The air should be
pumped out of the tube and as both balls will still reach the bottom of
the tube at the same time, this will dispose of the hypothesis that air
pressure causes objects to fall to the earth. If a small glass marble is
used it obviously disposes of the magnetism hypothesis also.

Experiment 13.5 (p. 67)

D.

Glass tube, 80 cm × 6 cm.
Rubber stoppers to fit (one should carry a piece of glass tubing
to which a rubber connection to an air pump can be attached).
Vacuum pump.

The above experiment can be repeated with a small coin or a washer
and a piece of paper or a feather in the tube. The effect of air friction

is obvious in the case of the paper or the feather. When the air is removed from the tube both objects fall together.

The effect of a force on motion

Good quality dynamics carts, roller skates, and large steel ball bearings will be found to travel surprisingly long distances when set in motion on smooth flat surfaces. On a carpet they soon come to a halt because of the rough surface. Reference should be made to riding a well-lubricated cycle along a smooth road in the absence of wind.

Experiment 13.6 (p. 68)

P.

Dynamics carts.
Large ball bearings.

Motion without friction

Experiment 13.7 and the following one are designed to show that in the absence of opposing forces a moving body will continue in steady motion in a straight line. It is difficult to get rid of friction in any laboratory experiment but modern techniques have made it possible to approximate to this condition. The linear air track or the air table is one method of achieving an almost frictionless motion. The principle of the method is that the moving body (often called a vehicle) rides on a cushion of air in much the same way as the Hovercraft. This is obtained by blowing air down a long tube of square cross-section in the sides of which are drilled a large number of small holes. Air is forced out of the jets. A vacuum cleaner working in reverse may be used to obtain the compressed air.

Another method of achieving this is to use a special 'puck' which moves over a surface on a cushion of gas. One of these pucks uses carbon dioxide, generated from solid carbon dioxide placed in the base of the puck. A very much simpler method is to use a 'balloon' puck. This can easily be made by the pupils. The cushion of air is supplied by a blown-up balloon attached to the puck which consists of a disc of wood or cardboard with a hole in it through which the air issues. The hole should be made very carefully to be exactly at the centre of gravity of the square or circular sheet. The hole should fit the bottom of the stopper exactly and the stopper (rubber or cork) should not be pushed so far into the hole that it extends beyond the surface of the disc. The balloon is inflated and attached to the tube passing through the stopper. This kind of puck works well on a smooth bench surface. One disadvantage of this puck is that the pupils might think that the balloon does more than provide a cushion of air on which the disc can travel; because they have seen a deflating balloon act like a rocket in which the issuing air provides the force for starting the motion they think that a similar thing is happening with the puck. The teacher should, however, discuss in which direction such a force would be acting, and what effect it would have on the motion of the puck. The pupils will soon find, if they are allowed to play with these pucks for a while, that the direction of motion depends on the initial push. They will discover that moving bodies continue to move unless a force is exerted to slow them down. This force, of course, does exist with the balloon puck and with the linear air track. There is always a certain amount of friction, although it has been reduced very considerably in these techniques.

Experiment 13.7 (p. 69)

D.

Linear air track apparatus with blower motor.

Experiment 13.8 (p. 69)

P. 10 groups.

10 balloons.

10 squares or discs of plywood, hardboard, or cardboard, drilled and provided with corks, etc. as P.T. Fig. 13.9.

For further information and guidance on the teaching of this section, teachers are recommended to read *Physics is Fun*, Book III, J. Jardine (Heinemann Educational Books Ltd.) and the relevant Teachers' Guide.

The balloon puck sheds some light on the fact that an orbiting spacecraft is able to continue in motion without the aid of any rocket thrust. It is moving in a vacuum where there is no friction. Other aspects of the motion of spacecraft are outside the bounds of this unit and should be left until later in the physics course.

Measuring force

Springs are preferable to elastic bands in making balances for the pupils to calibrate. The units of weight to be used will depend on the springs available but should be chosen so that a maximum of six of them can be added to the spring before the elastic limit is reached.

Experiment 13.9 (p. 70)
Stretching a spring

P. 10 groups.

10 stands.
10 springs.
10 sets of weights (e.g. slotted weights, identical washers, etc.).
10 white cards.
Graph paper.

The unit of force; the newton

The unit of force is the newton. It should not be defined at this stage as the definition depends on experiments to be performed later but, for the information of the teacher, the newton is that force which will give a mass of 1 kilogramme an acceleration of 1 metre per second per second.

It is important that the pupils should gain some idea of the size of this force. It is approximately the weight of a mass of 100 g. In the *Pupils' Textbook*, in order to give the pupils some conception of the newton in terms of everyday masses, it is quoted as being about the weight of a fair-sized apple but the teacher can probably devise equally relevant examples. For the teacher it may be worth while showing how this comes about. The acceleration due to gravity is 981 cm per second per second. The force of gravity exerted on a mass of 1 kg (i.e. the weight of 1 kg mass) causes it to accelerate as it falls to the earth by 9.81 m per second per second, or approximately 10 m s^{-2}. This force is therefore approximately 10 newtons, and 1 newton will be approximately the weight of a mass of 100 g.

Spring balances calibrated in newtons should be available. It is also useful to have a Nuffield *Forces Demonstration Box* which enables the pupils to feel a force of 1 newton without being able to see what they are lifting. This box is also useful in demonstrating the idea of work.

Experiment 13.10 (p. 71)
The newton spring balance

P. 10 groups.

10 spring balances calibrated in newtons.
A number of objects to weigh.

For demonstration purposes: 1 Nuffield *Forces Demonstration Box*.

Notes on the additional experiments (p. 71)

2. To find the number of newtons to which one of the pupils' units of force is equivalent, several of the units should be hung on a newton balance, and the weight of one of them calculated by division.

3. When the pupil adds more than a certain number of his units to the spring he was calibrating he should find that the elastic limit is exceeded and the graph levels off. Hence his spring would not be suitable for measuring forces of more than a certain value.

The lever

There are many different forms of apparatus suitable for this experiment. The one suggested in the *Pupils' Textbook* is a S.S.S.E.R.C. design. Various kinds of weight can be used. They need not have a marked value but might be identical large washers, metal discs, or metal squares.

In Experiment 13.11, Group 1, where the force is kept constant at a constant distance from the pivot on one arm, the pupils should find that the force and the distance of it from the pivot on the other arm required to balance are inversely proportional to each other. (The term 'inversely proportional' need not be used if the pupils have not encountered it in mathematics.) In Group 2 pupils should readily notice that when the lever is balanced the force multiplied by the distance from the pivot is the same on each arm.

Experiment 13.11 (p. 71)

P. 10 groups.

10 lever kits.
10 sets of weights (washers, discs, etc.).

In the *additional work* (p. 72) pupils should find that the greater the distance from the turning point or pivot that a force is applied, the less the force has to be in order to supply the necessary turning effect.

Pairs of forces

The series of experiments, grouped under Experiment 13.12, is best done in the form of a number of stations, as this affords maximum participation of the pupils with the minimum quantity of apparatus.

By the time the pupils have carried out each experiment in the series they will appreciate that in each case the forces have occurred in pairs, although, as in Experiment 13.12 and 13.13 only one of the pupils consciously exerted the force. In Experiment 13.12, if the water rocket is used in the classroom it should not contain water. The rocket will rise to the ceiling if only air is used. The pupils should be told the minimum number of strokes of the pump required; this should have been found by the teacher or technician before the lesson.

Another version of the water rocket which can be made in the laboratory is shown in Figure 13.1. An empty detergent bottle and a cycle tyre valve are used.

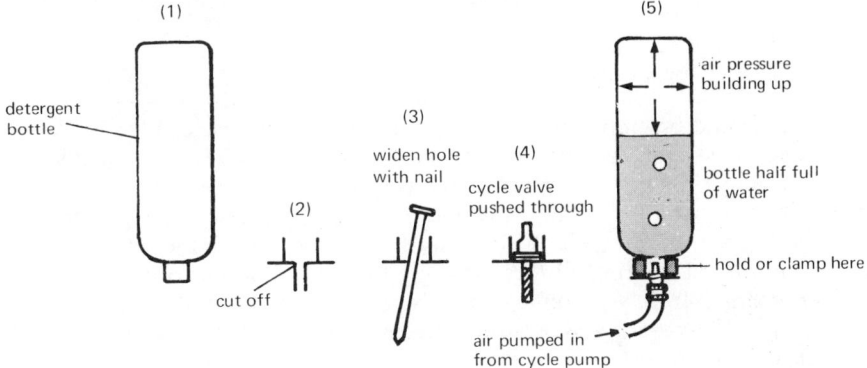

FIGURE 13.1

Pupils will enjoy having contests with these to find whose rocket rises to the greatest height. The explanation of the action of the rocket should be given in terms of pairs of forces.

Experiment 13.12 (p. 72)

S.

2 pairs of roller skates.
2 trolleys (such as apparatus trolleys).
1 pair dynamics carts.
Balloons.
Water rocket kit.
Friction drive car.
Polystyrene beads.
Paper.
Sellotape.
Thread.

Work and energy

It is important not to set out to define terms like work and energy. The pupils should instead be allowed to find the rule for calculating work done for themselves, as in the tables on p. 74 of the *Pupils' Textbook*.

Comparing the processes of lifting and dragging equal masses, the pupils should find that less force is required to drag an object than to lift it.

Experiment 13.13 (p. 75)
Work done in dragging

P. 10 groups.

Masses of various kinds.

Newton spring balances.

It should be emphasized to the pupils that in the case of lifting, the work done and the energy used have been transferred to potential energy. In the case of dragging, the energy used has been transformed into sound and heat. The same unit, the joule, is not only used for the work done but for all the various forms into which the energy is transformed. During all these changes there is an overall conservation of energy. It is very difficult to prove this experimentally with any degree of accuracy and it is not necessary to say much about it at this stage.

In Experiment 13.14, the pupils will each transport their own masses from one level in the building to a higher level, and so each will do work against the force of gravity by exerting a force equal to their own weight. This force is found by applying the fact already known to the pupils that the weight of a 1 kg mass can be taken as 10 N.

The distance involved is the vertical height of the stairs. The energy used will have been transformed into potential energy.

Each pupil should complete a table similar to that on p. 75 of the *Pupils' Textbook*. If possible the pupils should climb two stories in order to get reasonable differences between the powers of different pupils. The time taken should be measured to the nearest second.

Experiment 13.14 (p. 75)
Finding your own power

P.I.

Bathroom scales (calibrated in kg if possible; otherwise have a conversion scale available).

Stopwatch.

Measuring tape, or metre rule. The height of one step can be taken and multiplied by the total number of steps.

Units

As previously stated (*Teachers' Guide* to Book 1, p. 18) we are using SI units throughout this book. Pupils find these very straightforward to deal with, although teachers who have been accustomed to use British units or c.g.s. units may find them more difficult at first. Below is a summary of the SI units encountered in the course so far.

Quantity	Unit	Abbreviation
Mass	kilogramme	kg
Length	kilometre	km
Time	second	s
Force	newton	N
Weight	newton	N
Work done	newton metre	J
= Force × distance	or joule	
Energy	joule	J
Power	joule per second	$J\ s^{-1}$
= rate of doing work	or watt	or W
or rate of using energy		

It is permissible to use sub-units, such as the gramme, the metre, the centimetre etc.

Machines for lifting

We do not wish to have to take into account the weight of the lever itself in any calculations; to avoid this the lever should be balanced about its centre of gravity and in order to obtain meaningful results the rod used should not be uniform. The work done on the load will of course be the load (20 N) times the distance it moves (5 cm or 0.05 m), i.e. $20 \times 0.05 = 1$ N m or 1 J.

The work done by the effort will be the reading on the newton balance multiplied by the distance moved by the pupil's finger tips pulling on the balance while the load moves 5 cm.

Experiment 13.15 (p. 75)

P. 10 groups.

10 long levers.
10 newton spring balances.
10 2 kg masses.
10 metre sticks.

As an introduction to this lifting machine it is useful to discuss the action of systems of pulleys, such as are found in the home (Figure 13.2).

The advantage that pulley systems give in changing the direction of a force should be emphasized. It is generally easier to pull downwards than it is to lift a body.

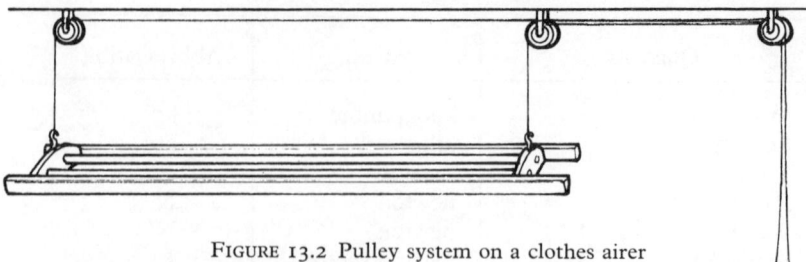

FIGURE 13.2 Pulley system on a clothes airer

Experiment 13.16 (p. 76)
The block and tackle

D.

Block and tackle sufficiently strong to support a pupil.
Strong rope for sling to support pupil.
Newton balance (0 to 200 N).
Selection of single and double pulley blocks.
Twine.

Support in plants

We now pass to consider forces in plants and animals and we start by dealing with methods of support in living things, beginning with aquatic plants, non-woody terrestrial plants, and woody terrestrial plants. The importance of water as a means of support can be seen by examining both aquatic plants and non-woody terrestrial plants. In the former the weight of the plant is partly supported by water. In water the aquatic plant remains upright; when it is removed from water it collapses. A woody land plant, however, remains erect because it has cells adapted for support.

Experiment 13.17 (p. 77)

P. 10 groups.

Plasticine.
Water weed—*Elodea*.
Woody twigs—privet or lime.
Beakers.

Photographs of transverse sections through the stems of an aquatic plant and of a woody twig are given in the *Pupils' Textbook*. They are there to show the obvious difference in the amount of supporting material in the stems. The pupils should make no attempt to learn the distribution of the tissues or to name them.

Experiment 13.18 (p. 78)

P. 5 groups.

Sunflower seedlings.

Flower pot or yoghurt container.

Vermiculite.

Test-tubes.

Labelling pencils.

For each group of four, plant four soaked sunflower seeds in vermiculite or soil in a flower pot or yoghurt container, 7–10 days before the lesson.

Water provides a form of internal support in seedlings. When the cells in the stem are turgid they form a relatively strong structure. If the seedlings are removed from water, water evaporates from the leaves. There is no water entering through the roots to replace this, the cells in the stem become flaccid, and the seedlings wilt through lack of support. If the seedlings are returned to water, water re-enters and they soon become erect.

Experiment 13.19 (p. 78)

(This is an additional experiment, but one which is done quantitatively.)

P. 5 groups.

Different kinds of seedlings, e.g. sunflower, pea, bean, etc.

Balance.

Oven set at 90° C.

Labelling pencils.

Evaporating basins.

Soaked seeds should be planted 7–10 days before the lesson. Allow 6–10 seedlings for each group. Each group can work with a different kind of plant and the results obtained from each can be used by all the class. After the initial weighing, the seedlings can be dried in an oven set at 90° C for 2–3 days. The percentage of water in each kind of seedling can be compared by presenting the results as a histogram.

Support in animals

The method of support in animals is also considered in a comparative way by examining a variety of invertebrates and vertebrates. The importance of water as a means of support in non-woody plants has been established in the last section. Water may therefore be a possible means of support in certain invertebrates.

Experiment 13.20 (p. 78)
A model worm

P. 10 groups.

10 'sausage' balloons.

Thread.

The above experiment can also be used to show the importance of body fluid in the movement of animals such as earthworms. It should be noted that this movement is only brought about by squeezing the balloon. The fingers represent the muscles in the body wall. This experiment should be referred to when dealing with muscles later (P.T. p. 83).

Other invertebrates, e.g. the arthropods, have additional support in the form of the exo-skeleton. This form of support can be shown by making a model arthropod. Cover an inflated balloon with a layer of papier mache. When the outside layer has hardened burst the balloon. The original shape is retained.

Experiment 13.21 (p. 78)

P. 10 groups.

10 'sausage' balloons.

Newspaper ⎱
Paste   ⎰ papier mache.

The comparative thickness or 'hardness' of exo-skeletons of members of the arthropods can be compared if a variety of animals are available. Locusts should be kept in the laboratory. They can be used in a number of experiments in the first two years of the course. Spiders, earwigs, woodlice, centipedes, etc. can be found under stones and in leaf litter in most gardens. They can be kept in the laboratory for several days in troughs containing leaf litter. (For the care of locusts see the notes on p. 84.)

Experiment 13.22 (p. 78)
External skeletons

P.

A variety of arthropods.

Support in vertebrates

The vertebrate skeleton is a firm supporting frame. It protects the internal organs, provides an anchorage for muscles and is essential for movement.

The relative sizes of land and aquatic vertebrates should be compared. Aquatic vertebrates can grow to a much larger size because their body weight is supported by water. Their skeletons are reduced. The skeleton of the whale is not large enough or strong enough to support the animal out of water.

The relative sizes of the pectoral and pelvic girdles of an aquatic mammal (the whale), a quadruped (the horse), and man should be compared. The sizes of the girdles can then be related to the method of locomotion of each animal. In the whale the girdles are greatly reduced. Movement is brought about by contractions of the body muscles and the tail. In the horse, weight is distributed on both sets of legs and both girdles are massive. In man, who is upright, the pelvic girdle, which supports most of the body weight is much larger than the pectoral.

Experiment 13.23 (p. 80)
Looking at bones

P. 5 groups.

5 specimens of femur bone of a bullock.
5 specimens of femur of a chicken.

The femur bone of a bullock can be obtained from a butcher. It should be cut lengthwise with an electric saw.

The femur of the bullock should be compared with that of the chicken to show the difference in size and in the nature of the bone itself.

Experiment 13.24 (p. 81)

P. 10 groups.

10 spring balances.
10 basins.
10 lumps of metal or pieces of bone.

The purpose of this experiment is to remind the pupils of the effect of buoyancy on the weight of an object. The object is weighed on the spring balance and then while still attached to the balance, it is lowered into a sink or basin of water.

Stability

Experiment 13.25 (p. 81)

P. 10 groups.

10 balls of plasticine each weighing 20 g.

Experiments 13.26 (p. 81) and 13.27 (p. 82)

P. 10 groups.

10 balls of plasticine (100 g).
Drinking straws.
10 pulleys.
10 clamps and stands.
String.
Balance pan or weight support.
Scissors.
Weights.

A model animal can be made using a plasticine cylinder for the body and straws for the legs (P.T. Figure 13.31, p. 81). The angle at which the straws are attached to the body can be altered, in effect altering the area of the base of the model.

Adding weights to a paper balance pan can be a slow procedure and the weights often fall off the pan. Two alternative methods for 'toppling' the model are illustrated below.

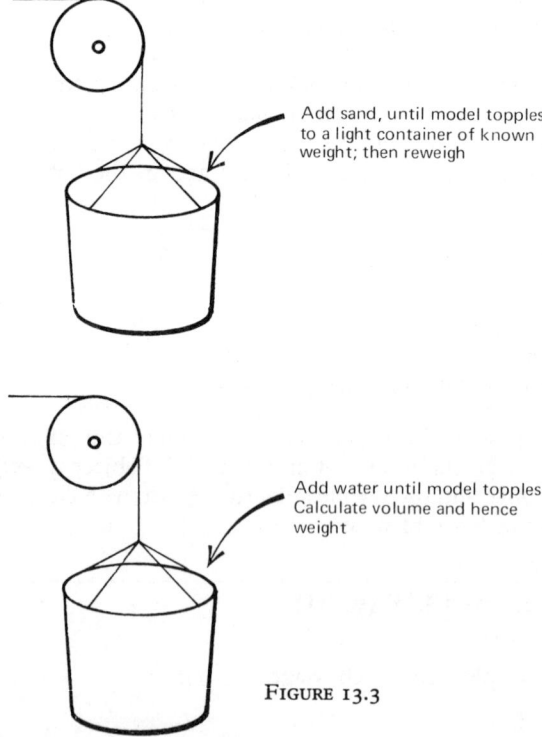

Add sand, until model topples, to a light container of known weight; then reweigh

Add water until model topples. Calculate volume and hence weight

FIGURE 13.3

If time permits, this work can be extended. The pupils can modify their most stable model, so that it will support a greater weight than that which originally toppled it. This activity can take the form of a competition to see which model will support the greatest weight. The models can be modified by adding up to 10 g of additional material to the model. This may be done by means of wire inserted through the straws, or plasticine feet, etc. The modifications, however, should be designed by the pupils themselves.

Body shape and stability

The centre of balance of a variety of 'animals' can be found, and compared, by using the shapes of the animals. It should be pointed out that the shapes represent only a slice through the animals.

Experiment 13.28 (p. 82)

P. 10 groups.

Cardboard.
Plumb line (a weighted string).
Drawing board, or piece of hardboard against which the cardboard 'silhouettes' can be supported.
Pins.

Muscles

This short section on muscles should be linked to the earlier work on levers. Pairs of antagonistic muscles, e.g. the biceps and triceps muscles move the bones in the limbs of our bodies. Muscles work by contracting. When the biceps contracts the lower part of the arm is raised and the triceps relaxes. To lower the arm the triceps contracts and the biceps relaxes. The importance and position of the tendons can be discussed at this point.

Experiment 13.30 (p. 83)
How much can your biceps muscles lift?

P. 10 groups.

A number of bricks.
Lengths of string.
Salter balance.
Pad of cloth.

A thick piece of cloth should be placed under the string before the bricks are lifted to avoid injury to the skin.

When an object is lifted by the forearm, work is done. The weight that can be lifted by the hand is less than the weight that can be lifted by the forearm. However, the distance moved by the former is greater than that lifted by the latter.

The arm acts as a lever. However, as the results of Experiment 13.30 will show, the effort exerted by the biceps is greater than the load.

Experiment 13.31 (p. 83)
What force is exerted by the biceps in lifting a load?

P. 10 groups.

10 metre sticks with hooks attached (as P.T. Figure 13.36, p. 84).
10 spring balances.
Weights.

Notes on the care of locusts in the laboratory

A cage suitable for keeping locusts in the laboratory is shown in Figure 13.4.

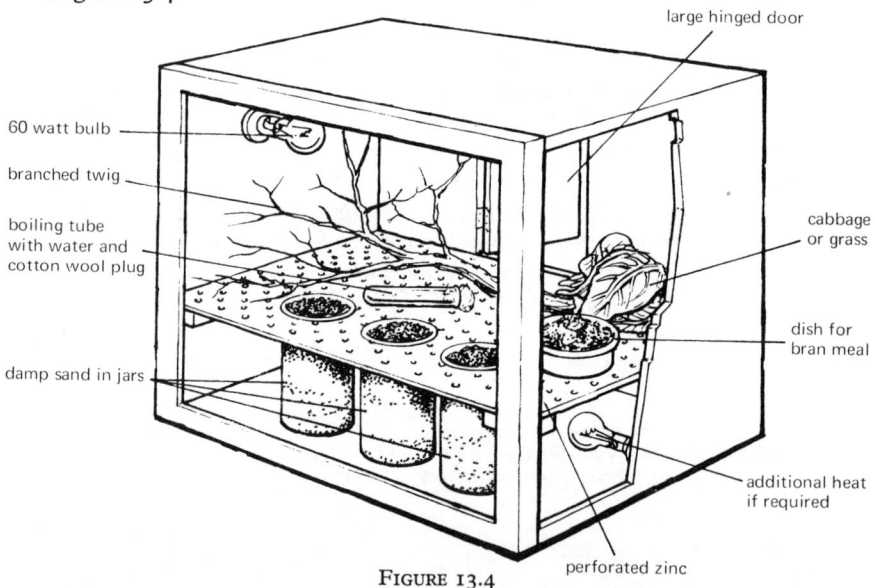

large hinged door

60 watt bulb

branched twig

boiling tube
with water and
cotton wool plug

cabbage
or grass

dish for
bran meal

damp sand in jars

additional heat
if required

FIGURE 13.4   perforated zinc

In order to breed locusts successfully, the temperature of the cage should be about 34° C during the day and 28° C at night. These temperatures can usually be achieved by placing a 60-watt bulb in the upper part of the cage and a 25-watt bulb at the base. By switching off the lower bulb the temperature will fall from 34° C to 28° C.

Locusts can be fed on a variety of foods. Grass, carrots, and cabbage are all suitable. This can be supplemented with bran. Water is not necessary if fresh food is used. Dried grass should be removed from the cage each day to avoid danger from fire.

Branches must be present in the cage. These are used as perches by the insects particularly during moulting.

Egg laying The female will lay eggs in the sand only if it is moist. To ensure the correct moisture content mix 100 parts by volume of sterile sand and 15 parts by volume of distilled water. The size of the sand particles can influence the development of the eggs. Several kinds of sand should be used to find the most suitable grain size. The sand should be heat sterilized.

After the eggs are laid in the sand, cover the tube, and no further water need be added. The tubes can be left in the cage, or put into an incubator. If the latter is preferred, label the tubes with the date on which the eggs were laid, and return them to the cage before the eggs hatch. They take 11 days to hatch at a temperature of 34° C.

Unit 14: Transport Systems in Living Things

Specific objectives

The objectives of this unit are that the pupils should acquire:

1. knowledge of some facts about foods and the means of classifying them,
2. knowledge of some facts about teeth,
3. some information about feeding in animals other than man,
4. knowledge of some facts about the digestive system of mammalia,
5. knowledge of some facts about various transport systems in plants and animals,
6. knowledge of some facts about elimination and excretion in plants and animals,
7. ability to apply knowledge to form classifications,
8. ability to relate structure to function,
9. ability to design experiments to obtain information from which to generalize, by investigating sweat excretion,
10. an interest in balancing food intake to ensure good health and proper body functioning,
11. an interest in maintaining healthy teeth,
12. awareness of the need for water balance in maintaining healthy animals and plants, and
13. further skill in simple biological techniques.

It is suggested that the *time allocation* for this unit should be approximately twenty-six periods.

Order of development

The topics covered in this unit can be divided into two main groups.

1. An investigation is made into food. This includes identifying food-stuffs, studying their uses, comparing the methods of feeding of different animals, and the development of the alimentary canal.

2. The need for transport systems is established. The development of transport systems is studied and the nature and function of such systems is compared in a variety of organisms. The problem of water balance in man is considered.

86

What are we talking about in this unit?

This unit follows on directly from Unit 5 (Book 1, pp. 89–91) where the process of digestion was studied. The tests for starch and glucose were established in that unit and these should be revised. Tests for proteins and fats are given in Experiment 14.1.

Experiment 14.1 (p. 86)
Testing for foodstuffs

P. 10 groups.

4 test-tubes per group.
10 beakers, bunsen burners, tripod stands and gauzes (or water baths).
10 labelling pencils.
10 dropping pipettes.
Filter paper.
Olive oil.
Starch solution (1 per cent).
Glucose solution (5 per cent).
Protein (egg white in suspension).
Millon's reagent.
Benedict's solution (or Fehling's solution).
Iodine.
(The reagents should be in dropping bottles.)

A suspension of egg white can be made as follows: separate the white of an egg from the yolk. Add a small quantity of the egg white to a beaker of very hot water. Stir the mixture quickly with a glass rod. A cloudy suspension should be formed.

Labelling test-tubes

A convenient way to label test-tubes which are to be heated in a water bath is to use tags which have been cut from old plastic detergent bottles.

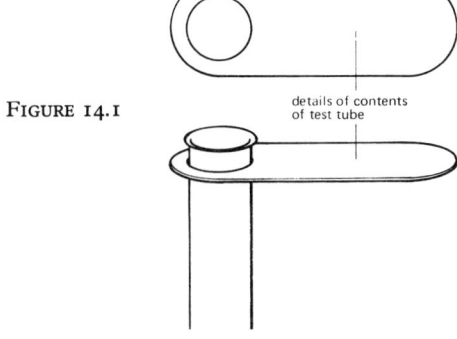

FIGURE 14.1

details of contents
of test tube

Experiment 14.2 (p. 86)
What substances are present in our food?

P. 10 groups.

4 test-tubes per group.
10 water baths.
Mortars and pestles.
Millon's reagent.
Benedict's solution (or Fehling's solution).
Iodine.
A variety of foods:
Foods rich in starch—potato, bread, flour
Foods rich in glucose—apple, grape
Foods rich in protein—cheese, meat, fish
Foods rich in fat—butter, lard.

Having established quick, sensitive tests for the main classes of food, the composite nature of foodstuffs can be investigated. Foods, such as fruit and vegetables, should be ground up before they are tested.

It should be remembered that sucrose itself is not a reducing sugar. When it is hydrolysed by boiling with dilute hydrochloric acid and the resulting liquid is neutralized with sodium hydroxide it responds to the reducing sugar test. Sucrose is hydrolysed to a mixture of glucose and fructose.

Experiment 14.3 (p. 87)
D.

Metal beaker.
Tin surrounded with asbestos.
Thermometer.
Glass rod.
Clay pipe.
Oxygen cylinder.
Rubber tubing.
Small bottle or boiling tube containing water which can be used to show the rate of oxygen flow from the cylinder.
Food dried in an oven between 90° and 100° C.

Figure 14.2 shows how to assemble the 'food calorimeter'.

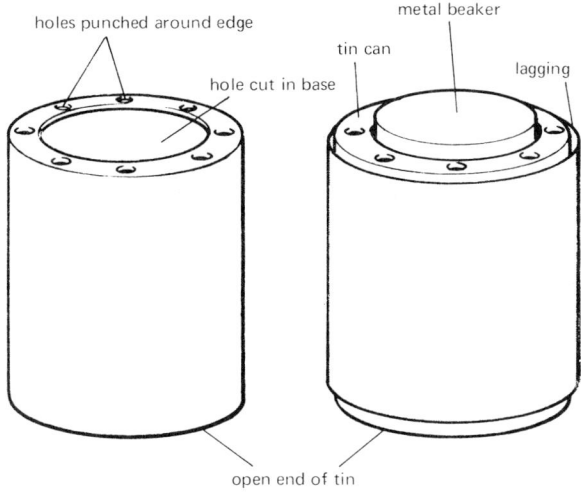

FIGURE 14.2

Details of the procedure for this experiment are given in the *Pupils'*
Textbook (p. 87). It is not designed to find the 'calorific value' of foods
but to compare the amount of heat released by a variety of foods. If
possible use similar foods to those tested in Experiment 14.2. The
foods must be dried in an oven at 90–100° C. The need for the removal
of water should be discussed and reference made to Experiment 13.19
where it was established that different seedlings contained different
amounts of water.

The results of this experiment should show that foods rich in
carbohydrates and fats give off most energy.

The importance of proteins, vitamins, and mineral salts is discussed
briefly in the text.

Feeding in animals

The structure and function of the four kinds of teeth are established.
If extracted teeth are available, cut them longitudinally, smooth the

cut surface and examine the internal structure. If this is not possible, the structure can be shown in diagrams and coloured charts of teeth. Charts can be obtained free of charge from the General Dental Council, 137 Wimpole Street, London, W.1.

Experiment 14.4 (p. 88)

I.

Any piece of fruit—apple or pear, for instance.
Bar of toffee.
The materials for this experiment need not be provided by the teacher!

Care of the teeth

A survey of the amount of dental caries amongst members of the class should show the need for care of the teeth. The survey can be carried out with all classes and the results presented in the form of a histogram.

Experiments 14.5 and 14.6 attempt to show how tooth decay arises and Experiment 14.7 shows the part played by tooth paste in helping to prevent tooth decay. This is two-fold. The paste is alkaline in nature and thus helps to neutralize the acid in the mouth. However, since the paste is in the mouth for a comparatively short time this is not the important factor. Tooth paste contains particles which act as a very fine abrasive and help to scrape bacteria from the surface of the teeth. Brushing the teeth with salt or soot would have a similar effect.

Experiment 14.5 (p. 89)

P. 10 groups.

Teeth. (Sheeps' teeth can be used in this experiment.)
Wax.
Needle.
Crystallizing dish.
Dilute hydrochloric acid.

Experiment 14.6 (p. 89)

P.I. or 10 groups.

Test-tubes, 4 for each group.
Labelling pencils.
Spotting tiles.
pH paper or litmus paper.
Dropping pipette.
1 per cent glucose solution.

Experiment 14.7 (p. 89)

P.I. or 10 groups.

Test-tubes.
pH paper or litmus paper.
Dropping pipettes.
Microscope slides.
Tooth paste.
Microscopes.
Cover slips.

Experiment 14.8 (p. 89)

Apparatus as in Experiment 14.7 but include a variety of tooth pastes.

Teeth in other mammals

The structure of teeth and the jaw action of a herbivore (sheep) and a carnivore (dog) can be compared and related to the feeding habits of each animal (Experiment 14.9, p. 91).

If possible, the skulls of the animals should be examined. The shape of the molars can be linked to the direction in which the jaws move. In the herbivore the side to side action of the jaws, and the curved ridges on the teeth, ensure that the food is well ground. The grinding process can be compared with the milling of wheat. This is important because the plant material which is eaten is rich in cellulose and can be very tough.

Herbivores do not have to catch their prey and so the canine teeth are greatly reduced or absent.

In carnivores most of the teeth have narrow, cutting edges and the jaws move up and down like the blades of scissors. The canines are well developed for catching and holding food. The food is bitten into small pieces and then swallowed after little, if any, chewing.

Feeding in invertebrates

Some of the different methods of feeding in invertebrates can be compared by examining the locust, the housefly, and, if possible, the mussel.

Experiment 14.10 (p. 92)

P. 10 groups.

Jam jars.
Muslin to cover.
Locusts.

> Jar *or* boiling tube plugged with cotton wool.
> Hand lens.

Instructions for the care of locusts in the laboratory are given on p. 84. When the adult locusts die they can be stored in alcohol and then dissected to show the mouthparts. If enough locusts are available they can be dissected by the pupils; if not, the dissection can be demonstrated.

Dissection of mouthparts of locust

1. Hold the locust with the head between the points of a pair of blunt forceps.
2. With a pin lift the upper lip. Remove this by pulling it gently with a pair of fine forceps. Lay the upper lip on a slide. The hard, black, shiny mandibles can now be seen.
3. Insert the point of the forceps between the mandibles and force them apart to show the serrated inner edge.
4. Pull off the mandibles and place them on the slide at the side of the upper lip. (Note: if preserved material is used, hold the head firmly before removing the mandibles or the head may become separated from the body.)
5. Remove the lower jaws and place these at each side of the mandibles.
6. Remove the lower lip and palps and place it on the slide.

The mouthparts should now be displayed on the slide. The 800 E film loop *The locust as a biting insect* (N.B.P.—6; Longmans) shows how the various mouthparts are used in feeding.

> ### Experiment 14.11 (p. 93)
> P. 5 groups.
>
> Locusts.
> Supply of grass.
> Balances.

This is an additional experiment, but one which can be used to illustrate the magnitude of the problem of locusts. The experiment is not accurate because no allowance is made for the loss of weight due to the evaporation of water from the grass. This point should be discussed with the class. Another possible approach to this experiment is to find the weight of a locust and then, using the fact that a locust eats its own weight of food per day, calculate the weight of food eaten by a swarm in a day.

> ### Experiment 14.12 (p. 93)
> P. 10 groups.

Jam jars.
Muslin to cover.
Elastic bands.
Hand lenses.
1 per cent glucose solution.
The 800 E film loop *The fly as a sucking insect* (N.B.P.—7; Longmans) should be shown, if possible.

The mussel as a filter feeder

If freshwater mussels are available they can be used to demonstrate filter feeding. They can be kept in a tank in the laboratory. The bottom of the tank should be covered with sand. These mussels eat dead organic material found in the water. This can be provided by shaking pond weed into the water.

Mytilus should only be used for this experiment if it is obtained from a firm of biological suppliers or collected from a shore free from pollution by sewage. Because it is a filter feeder, high concentrations of toxins can gather in the mussel.

Experiment 14.13 (p. 93)

P. 10 groups.
Binocular microscopes or large hand lenses.
Microscopes.
Slides.
Mussel with one valve removed.
Piece of gill removed from a second mussel.
Fine carbon or carmine particles.

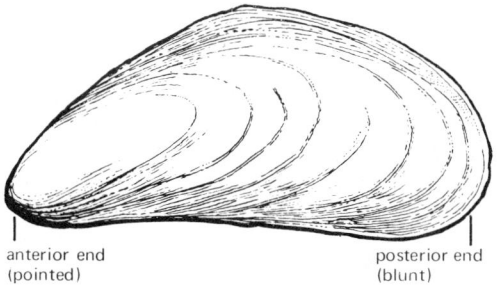

anterior end   posterior end
(pointed)   (blunt)

FIGURE 14.3 A mussel, *Mytilus*

To open *Mytilus* prise the valves slightly apart with a blunt scalpel. The valves are held together by two adductor muscles, one at the anterior (pointed) end and one at the posterior (blunt) end. To remove a valve these muscles must be cut. Insert a sharp scalpel between the valves at the anterior end and cut through the adductor muscle. Pull the valves apart and place the mussel in a dish of sea water. The gill flaps should now be visible. Allow the mussel to recover for 5–10 minutes. Open a second mussel and cut a small piece of tissue from the gills. Mount this in salt water on a slide. Examination of this under the low power of the microscope will show the cilia. Mud particles placed on the gills of the opened mussel are carried by the beating of these cilia. Figure 14.4 shows the direction of the movement of the mud particles.

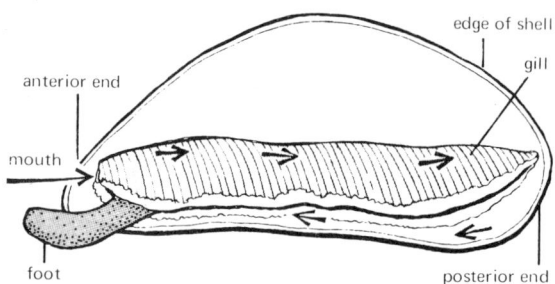

FIGURE 14.4 *Mytilus* opened to show direction of flow of mud particles on cilia

Digestive systems

The structure of the alimentary canal of *Daphnia* or a nematode worm, an earthworm, and a mammal is studied to show that each is basically a tube, open at each end. In the larger organisms parts of this tube have become modified to perform a particular function, e.g. in the earthworm the gizzard, where food is ground up, is hard and muscular.

Experiment 14.14 (p. 93)

P. 10 groups.

Microscopes.
Microscope slides.
Cover slips.
Dropping pipettes.
Mounted needles.
Pond water containing *Daphnia*.
or white nematode worms.

White nematode worms can be found in freshly dug soil. A culture of them can be maintained easily in the laboratory by adding a piece of bread occasionally to a sample of damp garden soil.

Microscopic examination of either of these animals should show clearly the movement of the walls of the intestine. This movement is independent of the movement of the body wall.

Experiment 14.15 (p. 94)

D or P.

Apparatus for dissection: forceps, scalpel, scissors, needle, and pins.
Pie dish or small enamel dish filled with wax.
Large earthworm.

The earthworm can be killed by dropping it into a beaker of boiling water or placing it in a 'killing chamber'. A desiccator can be used for this purpose if cotton wool soaked in ether is placed in the lower chamber. The worm should be killed before the class arrives.

Dissection of an earthworm

1. Fill a pie dish a quarter full with molten wax. Allow the wax to harden.
2. Place a large earthworm on the wax so that the dorsal surface is uppermost. The dorsal surface is brownish-red in colour; the ventral surface is paler.
3. Insert a pin through the head and one through the tail end.
4. Using a scalpel or fine scissors make a cut through the dorsal surface from the head to a point below the clitellum. There is no need to cut the entire length of the worm.
5. Using fine forceps, lift back one side of the skin and break the rings of tissue between each segment (the septa) by running a needle below the skin. Pin back the skin with fine pins. Pin back the other side in the same way. The alimentary canal should now be visible.
6. The creamy coloured structures lying between the pharynx and the crop are reproductive organs. Remove these to expose the oesophagus.

Experiment 14.16 (p. 94)

D.

Dissecting instruments.
Large dissecting pins.
Dissecting board.
Rat.

Dissection of a rat

N.B. Rats used for dissection must be obtained from a biological supplier or must be bred from rats obtained from such sources.

1. Lay the rat on a dissection board with the ventral surface uppermost.
2. Insert pins through the feet.
3. Lift the skin in the abdominal area with blunt forceps and make a small cut through the skin.
4. Insert the point of a pair of fine scissors through the cut and cut the skin up to the pectoral girdle and down to the urethra.
5. Cut the skin along each limb.
6. Lift the edge of the skin and separate the skin from the muscular abdominal wall by running a scalpel between them.
7. Pin back the skin.
8. Cut the abdominal wall as in stages 3 and 4.
9. Pin back the abdominal wall. The alimentary canal should now be visible. The stomach, small intestine, and caecum can be seen.
10. To display the small intestine, caecum, and large intestine, unravel the loops of the small intestine. The intestine is twisted into loops but it can be unravelled by pulling it gently with blunt forceps and cutting through the tough mesentery between the loops. Pin out the intestine as it is unravelled. The blood vessels in the mesentery should be noted before it is cut.
11. Remove the alimentary canal by cutting through the oesophagus above the stomach, and the rectum.

Transport systems

The need for transport systems should be discussed. The cell as an organism or as part of a large complex organism requires food, water, and oxygen for its chemical activities. Waste products resulting from these activities must be removed.

In unicellular organisms, which have a large surface as compared to their volume, these substances diffuse through the cell membrane. However, larger multicellular organisms have a smaller surface area compared to their volume, and the diffusion of food and other products into the body would be much too slow to maintain life. Certain regions of the body have become adapted for the absorption of one particular substance, e.g. air is absorbed through the lungs and digested food through the small intestine. The substances are then transported to all parts of the body.

Transport systems in plants

Experiment 14.17 (p. 95)

P. 10 groups.

10 scalpels or knives.
10 hand lenses.
10 beakers.
10 slides.
10 cover slips.
Microscopes.
Plants, such as 'Busy Lizzie', celery, Brussel sprout.
Whole plants such as geranium, groundsel.

To ensure that the xylem cells are completely stained with eosin allow the stems to stand overnight in the dye.

Thin sections of stem can be cut with a razor blade. Moisten the blade with water and draw it several times over the cut end of the stem. Transfer the sections to water in a watch glass and mount the thinnest ones on a slide.

The xylem cells will be stained red. If these tubes are removed from a longitudinal section of the stem and examined under the microscope, the bands of strengthening material can be seen.

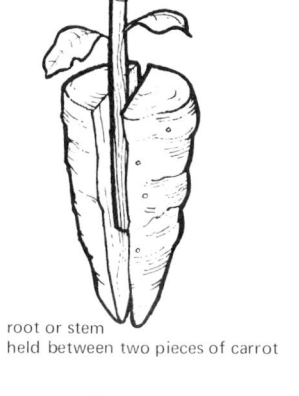

root or stem
held between two pieces of carrot

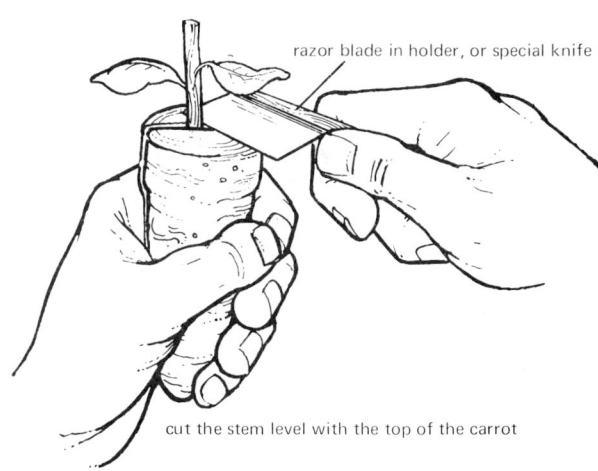

razor blade in holder, or special knife

cut the stem level with the top of the carrot

FIGURE 14.5 (continued on p. 98)

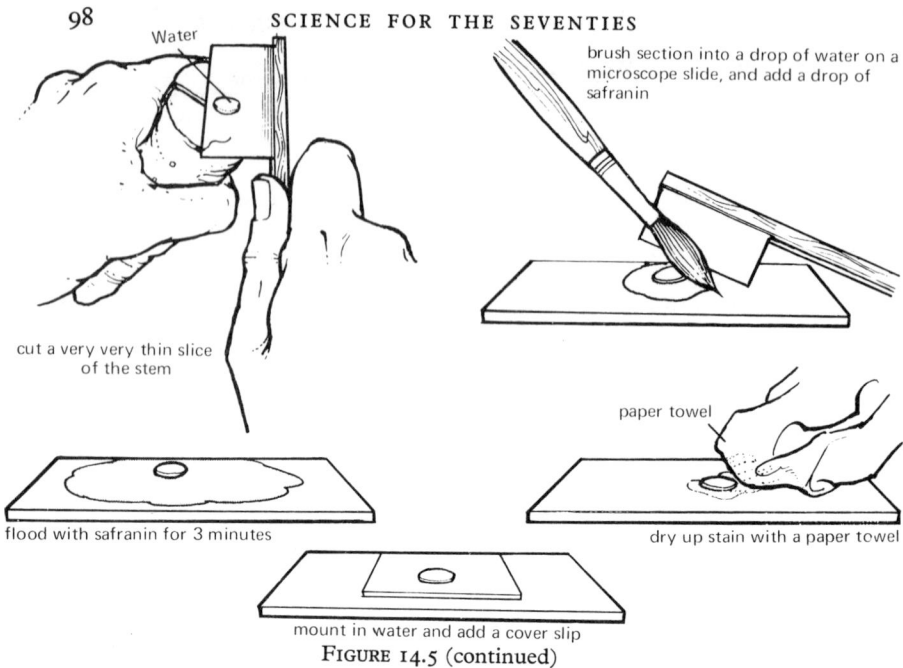

Water

brush section into a drop of water on a microscope slide, and add a drop of safranin

cut a very very thin slice of the stem

paper towel

flood with safranin for 3 minutes

dry up stain with a paper towel

mount in water and add a cover slip

FIGURE 14.5 (continued)

Experiment 14.18 (p. 97)

D.

Beaker.
Stop clock.
Air blower.
Stems of 'Busy Lizzie' or celery.
Eosin.

Transport systems in animals

In animals there are three basic types of transport systems:
1. The simple type as exhibited by cytoplasmic streaming in *Paramecium* (or cells in the leaf of Canadian Pond Weed).
2. The open type as seen in *Gammarus* and *Daphnia*.
3. The closed type as seen in earthworm, fish, and other higher animals.

Experiment 14.19 (p. 97)

D.

Microscope.
Microscope slides.
Cover slips.

Canadian pond weed.
Paramecium.
Yeast cells stained with congo red.

Yeast can be stained with congo red as follows: add 0.5 g congo red to 250 cm³ distilled water. Add 5 g of powdered yeast. Boil for 15 minutes and allow to cool. Place a drop of culture of *Paramecium* on to a slide and then add a drop of stained yeast cells. After about 3 or 4 minutes add a drop of methyl cellulose (Polycell) to slow down movement. Do not add the methyl cellulose before this because it prevents yeast cells entering the *Paramecium*. Cover with a cover slip and set up under the microscope. Large species of *Paramecium*, such as *Paramecium caudatum*, should be used in this experiment.

Details for the preparation of leaves of Canadian pond weed are given in *Teachers' Guide* to Book 1, p. 165.

Experiment 14.20 (p. 98)
D.

Microscope.
Slide.
Wide mouth pipette.
Gammarus or *Daphnia.*

Experiment 14.21 (p. 98)
D.

Microscope.
Slide.
Cotton wool.
Goldfish or *Xenopus* tadpole.

The tail of a goldfish can be examined under the microscope as follows: soak a pad of cotton wool in the water in which the fish is living and squeeze out excess water. Remove the fish and wrap the wet cotton wool around the gills at the head end. Hold the fish firmly but gently so that the tail is under the low power of the microscope. Blood vessels and blood should be clearly visible. Blood can also be seen in *Xenopus* tadpoles and trout alevins.

The heart

Experiment 14.22 (p. 99)
D.

Dissection instruments.
Dissection board.
Sheep's heart.

Dissect the heart by first cutting through the side of the right auricle (atrium) and the right ventricle. This simple dissection shows the thin auricle wall, the thick muscular wall of the ventricle, the tricuspid valve and tendons, and the blood vessels entering and leaving the heart. A second cut can then be made through the corresponding chambers in the left side of the heart.

Experiment 14.23 (p. 99)

P. 10 groups.

Piece of wood, e.g. ruler or meter stick.

Experiment 14.24 (p. 99)

P. 10 groups.

Stop clocks.
Graph paper.

Experiment 14.25 (p. 99)

P. 10 groups.

Test-tubes.
Measuring cylinders.
Straws.
Graph paper.
Stop clocks.

Experiment 14.26 (p. 100)

P. 10 groups.

As for Experiment 14.24.

Experiment 14.27 (p. 100)

P. 10 groups.

Centrifuge.
Dropping pipettes.
Test-tubes.
Water baths.
Bullock's blood (from slaughter house).
Millon's reagent
Benedict's solution (or Fehling's solution) } in bottles with dropping pipettes.
Iodine.

Blood and waste substances

> **Experiment 14.28 (p. 100)**
> D.
> Dissection instruments.
> Kidney.

Dissection of a kidney

The kidneys of the sheep or pig are suitable for this dissection because of their size. When ordering the kidney ask that the tubes attached to it, i.e. the blood vessels and the ureter, be not cut too close to the body of the kidney.

The kidney should be cut lengthwise by running a sharp blade around the convex side. When it is opened, the cortex, medulla, pelvis, and ureter can be clearly seen.

> **Experiment 14.29 (p. 101)**
> P. 10 groups.
> 10 measuring cylinders.
> Milk bottles.
> Cups.
>
> **Experiment 14.30 (p. 101)**
> P. 10 groups.
> Filter paper.
> Adhesive tape.
> Balances.
> Polythene.
>
> **Experiment 14.31 (p. 101)**
> P. I.
> Ether.
> Dropping pipette.
> N.B. Ether is very inflammable, and there must be no flames in the laboratory during this experiment.

Homeostasis

The fact that the average amount of water in the body remains fairly constant can be used to introduce the idea of homeostasis. There should be no attempt to use or learn the term homeostasis but the idea of the maintenance of a steady internal environment should be introduced.

The body responds to changes in its internal environment. If the amount of water in the body drops, small amounts of concentrated urine are produced and a sensation of thirst encourages drinking.

Sweat

The chief function of sweat is to produce a reduction in body temperature. If water is not needed for this purpose it is excreted in urine.

Bird pellets

If the pellets of flesh-eating birds are available, they can be dissected. These birds can separate flesh from the fur and bones of their prey. The flesh is digested and the undigested material is regurgitated.

Waste products in plants

Plants do not possess well developed excretory systems. Waste products are often deposited in regions of the plant that are likely to be discarded, e.g. the leaves and fruits. When these fall from the plant the waste material is removed.

Unit 15 Electricity and Magnetism

Specific objectives

The objectives of this unit are that pupils should acquire:

1. some information about the relationship between electrical units,
2. some information about costing electrical energy,
3. knowledge of the use of beam deflection in a cathode ray tube,
4. knowledge of some facts about discharge tubes,
5. knowledge of some facts about electromagnetism,
6. knowledge of some facts about the motor-effect and its application,
7. knowledge that a current can be generated by relative motion of a closed coil and a magnetic field,
8. knowledge that there is alternating current as well as direct current,
9. ability to apply knowledge of electrical circuitry to domestic wiring,
10. ability to analyse current relationships in parallel circuits,
11. ability to calculate fuse values for given situations,
12. awareness of the important technological revolutions resulting from the development of electromagnetism and the later development of electronics,
13. awareness of, and interest in, leisure pursuits in electronics, and
14. further skill in wiring techniques.

It is suggested that the time allocation for this unit should be about twenty-eight periods.

Introduction

The teacher's treatment of this unit will depend very much on the type of class he is presented with. If the class is composed entirely of boys interest will easily be maintained even with sets of poorer ability. For classes composed entirely of girls the teacher may perhaps have to be more selective and deal with the topic less deeply. With mixed classes of boys and girls the enthusiasm of the boys will probably carry the girls' section along. If this unit is being taught by a non-specialist teacher it would be helpful if the assistance of a physics specialist were obtained when rehearsing the demonstration of special apparatus and kits which are not very familiar.

Safety and danger with electricity

It is absolutely imperative that at the outset of any work on electricity the dangers of pupils interfering with mains and appliances at home be spelt out. Pupils becoming familiar with low-voltage supplies connected to the mains should not be allowed to assume that it is safe for them to experiment with mains supplies at home.

This unit starts off with the pupils discovering which insulators are in common use in appliances in the home. Kits of a large number of materials, both insulators and conductors, should be available for the pupils to test. A torch bulb can be used as an indicator of current flow.

Circuit boards will be used, and it is advisable for the teacher to read the *Teachers' Guide* to Book 1, pp. 143 ff, where this equipment is more fully discussed.

Experiment 15.1 (p. 103)
Conductors and insulators

P. 10 groups.

10 circuit boards and components (see below).
10 kits containing
    (a) strips of insulators: rubber, bakelite, glass, porcelain, mica, cotton, perspex, PVC, other plastics, etc.
    (b) strips of conductors: copper, brass, lead, iron, carbon, aluminium.

Components Each circuit board should be equipped with at least the following:

3 U2 cells.
6 connector links.
2 links with crocodile clips.
2 crocodile clips.
1 switch (link).
1 rheostat, 10 ohms.
4 bulb holders, fitted as links.
4 2.5 V bulbs.
1 length of 32 s.w.g. nichrome wire.
Wire wool.

It is useful to have an assortment of electrical components, e.g. switches, fuse holders, plugs, sockets, wires and cables, heating elements, where the pupils can see the insulators and name them.

Electricity in the home

Experiment 15.2 (p. 103)

P. 10 groups.

10 circuit boards with components.

10 ammeters (0–500 mA or 0–1 A).

Switching the current in and out of each branch of the parallel circuit in P.T. Figure 15.2 can easily be effected by putting in or taking out links; the links with crocodile clips and the switch link can be employed here.

At this stage it is useful to show a two-dimensional lay-out of the wiring of a model house—say the outline of a house in hardboard using small lampholders, 3.5 V bulbs and miniature switches for lights, and 3.5 V bulbs inside match boxes etc. to simulate TV, electric fires, cookers, etc. The house may have about four rooms and could be powered from a 4.5 V dry battery (3 U2 cells, for instance) or a 4 V accumulator. The model can be assembled as an after-school project or made by the science club.

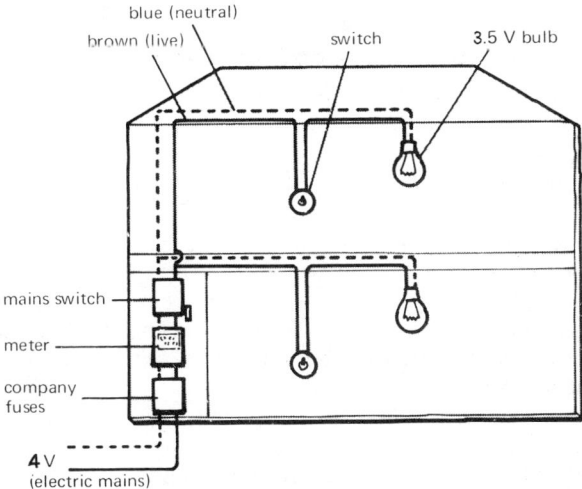

FIGURE 15.1 Model on hardboard of domestic electric wiring system

Wall charts and literature available from the Electrical Association for Women will be found very useful in dealing with this section.

Measuring electrical energy

Various electricity meters can be obtained for demonstration use in school and it is handy to have one which can be opened up to show the mechanism. The pupils must first be warned that it is illegal to do this with the electricity meter installed in their home! Pupils should be shown the gearing system in the dial type in which the hands on adjacent dials rotate in opposite directions. This should be done only while the meter is disconnected from the mains.

It is useful to have a hardboard model of the dials of such a meter so that pupils can have practice in reading them. This could be done round the class, or by inviting the pupils to act as meter readers. They could mark up the readings on the blackboard.

The rule for reading the dial type meter is to take the number the pointer has just left.

In the interests of safety it is not advisable to show a kilowatt-hour meter working on mains voltage. Joulemeters are available which work on 12 V alternating current. They can be shown working while the current is passing through a car headlamp or a low voltage immersion heater (such as the Nuffield type heater) placed in a beaker of cold water.

N.B. The joulemeter must be used only on 12 V alternating current.

Experiment 15.3 (p. 104)
Experiments with an electricity meter

D/P. The number of groups will depend on the equipment available.

Low-voltage power packs, a.c. output 12 V.
Joulemeters.
12 V 48 W car headlamps in holders; or 50 W low-voltage immersion heaters.
12 V 6 W car sidelamps in holders.
Stop clocks.

When the current is switched on through the bulb the hand recording hundreds of joules will be seen to rotate; faster with the headlamp than with the side lamp.

The joulemeter should be read and the current then switched on for 100 seconds (or 200 seconds) and then read again at the end of this time. The difference in readings in 100 seconds for a 48 watt lamp will be 4800 joules (1 joule is 1 watt second).

Pupils could use lamps of different wattages and find the number of watts used by dividing the meter reading by the time in seconds for which the current flows. The result could be compared with the stated wattage of the lamp.

For measuring electrical energy at home the kilowatt hour is used. 1 kilowatt hour is clearly 1000 watts for 3600 seconds, i.e. 3 600 000 J.

Answers to the examples in Section 15.4

(a) No. of units = power in kW × time in hours
$$= 3 \times 10$$
$$= 30 \text{ kWh or 30 units}$$
Cost at 1p per unit = 30p.

(b) Suppose there are 30 days in the month.
No. of units = 0.15 × 4 × 30
$$= 18 \text{ units}$$
Cost at 1p per unit = 18p.

(c) No. of units = 0.005 × 24 × 365
$$= 43.8 \text{ units}$$
Cost at 1p per unit = 44p.

Experiment 15.4 (p. 105)
D.

Lamps of various powers—60, 75, and 100 W.
Demonstration A.C. ammeter 0–1 A.
Demonstration A.C. voltmeter to cover mains potential.

The pupils should be asked to fill in the blanks in the table (P.T. p. 106). Further examples on these lines can be formulated using the power ratings of everyday appliances.

It should be emphasized to the pupils that a 13 A fuse gives very little safety at all where much lower currents are involved and that when a so-called 13 A plug is purchased, the 13 A fuse should be changed for a fuse of the value appropriate to its purpose.

Electronics or electrons in motion

Experiment 15.5 (p. 106)
Charges

P. 10 groups.

It is not necessary to supply complete electrostatics kits. The following articles only are required:

10 gold-leaf electroscopes.
10 strips of clear cellulose acetate.
10 strips of polystyrene.
10 dusters.

In place of the plastic materials listed, it is, of course, possible to use the rods which have traditionally been used in electrostatic experiments —glass, sulphur, ebonite, etc. Other materials will probably come to hand, such as strips cut from empty detergent bottles. It is important that of the pair of materials chosen one should become positively charged when rubbed with the duster and the other negatively. If the materials suggested above are not used the type of charge will depend on the material for rubbing—silk, fur, etc.

It is suggested that the Van de Graaff generator used should be one which has a 4 mm hole drilled in the dome, so that leaves made of balsa wood can be inserted, as in Figure 15.2.

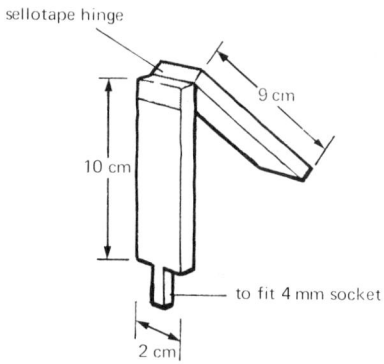

FIGURE 15.2 Balsa wood leaves for a Van de Graaff dome

They are hinged together with sellotape.

The sensitive meter should be a micro-ammeter. The current obtained will depend on the humidity of the air. For electrostatics experiments it is best to choose a dry day and have the room well ventilated. Pupils should not cluster too closely round the Van de Graaff generator. If using a scalamp galvanometer switch to the least sensitive range initially and adjust later to get the best reading on the scale.

Experiment 15.6 (p. 107)
Cancelling out charges

D.

Nichrome or contra wire spiral.
Low-voltage power supply.
Van de Graaff generator.

The spiral (P.T. Figure 15.6) should be of 20–30 cm of thin (about 32 s.w.g.) nichrome or contra wire mounted on an asbestos strip. The voltage used to make the wire red hot will depend on the length used but will be of the order of 12–15 V.

Electrons given off by a heated wire are sometimes referred to as 'thermions' in electronics and explain the action of thermionic valves in radio, etc. These are now quickly giving way to transistor devices but valves are much more easily understood by pupils at this stage.

One-way conduction

Experiment 15.7 (p. 107)

Demonstration diode valve.
E.H.T. supply.
Milliammeter.
Low-voltage supply for filament.

Demonstration diodes are expensive and require care in handling. They are best mounted in special stands. Care should be taken not to apply more than the permitted maximum voltage to the heater—this is probably 6 V. Care should be exercised with the E.H.T. supply and pupils should not be allowed to point fingers towards connectors at high potential.

It is interesting for the pupils to identify the electrodes in E.H.T. diodes from old TV receivers; here the anode is thimble-like and the heater is the cathode.

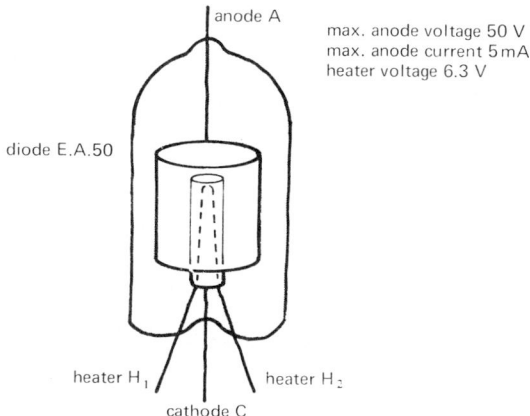

FIGURE 15.3 The valve EF 98 can be used alternatively with the anode and grid connected together. Maximum anode voltage 12 V; heater voltage 6.3 V.

If demonstration diodes are not available it is possible to use E.A.50 diodes or their equivalents, which should be wired as shown in Figure 15.4.

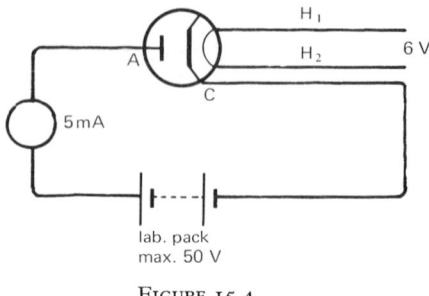

FIGURE 15.4

Alternatively many thermionic valves can be used with the grids and anodes connected together.

Transistor diodes

Experiment 15.8 (p. 108)

P. 10 groups.

10 circuit boards with components.
10 transistors.
10 10 ohm carbon resistances.

Many transistor diodes (e.g. P.T. Fig. 15.8) will be suitable for use with pupil circuit boards and voltages of 3 V. One such is OA 81. Transistor diodes are often marked red at the end to be connected towards the positive terminal of the battery supply for conduction to be possible. If reversed the pupils will not detect conduction using a torch bulb as indicator.

The diode should be replaced in the circuit board by a 10 ohm carbon resistance, which, of course, does not have directional properties. Other low value carbon resistances (e.g. 25 ohms) can also be used.

The colour code to be found on old carbon resistances is as follows:

Colour	BLACK	BROWN	RED	ORANGE	YELLOW
Number	0	1	2	3	4
Mnemonic	Bad	boys	run	over	young

Colour	GREEN	BLUE	VIOLET	GREY	WHITE
Number	5	6	7	8	9
Mnemonic	grass	but	violets	grow	wild

i.e. black represents the number 0, brown the number 1 and so on. The third coloured band represents the number of zeros after the first two figures. Thus a 10 ohm resistance would have one brown coloured band and 2 black. A 25 ohm resistance would have coloured bands red green, and black.

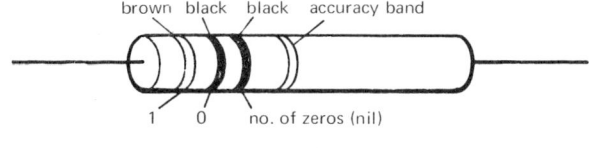

value of resistor = 10 Ω

FIGURE 15.5

Experiment 15.9 The Maltese Cross Tube (p. 108) is not essential in the course but adds considerably to the pupil's interest, as its operation helps in understanding the cathode ray tube used in TV sets.

Again, great care should be exercised in handling expensive apparatus of this kind. If the cathode only is switched on (care should be taken not to exceed the permitted heater voltage) the cathode will glow with a light which is often strong enough to throw a shadow of the Maltese Cross on the screen. Pupils should be asked about the significance of this and be reminded that light rays travel in straight lines (Unit 11). If the E.H.T., connected as shown (P.T. Fig. 15.10), is switched on, the fluorescent screen will glow but again there will be a shadow of the Maltese Cross. This occurs because the beam of electrons which is 'boiled-off' the cathode is accelerated through the hole in the anode and collide with the screen particles with sufficient energy to cause them to fluoresce. Like light rays, these electrons must travel in straight lines to explain the sharpness of the shadow produced. Pupils should be allowed to see the effect of increasing the E.H.T. voltage to 5000 V; this speeds up the electrons giving them more kinetic energy which is converted into further light energy so that the screen appears brighter. When a strong magnet is brought near the screen the illuminated area will be deflected perpendicularly to the direction of approach of the magnetic pole.

If a pupil type c.r.o. is again shown to the class they could be asked to suggest how the brilliance and shift controls function. The anode voltage could control the brightness and a magnetic or, more likely, an electrostatic field could control the movement of the electron beam which shows as a spot where the electrons hit the screen.

Note It is no longer considered safe to use 'cold-cathode' Maltese Cross Tubes, i.e. those which are not provided with a heated cathode. The maximum E.H.T. voltage permitted is 5000 V in case X-rays are generated.

Strip lighting

Experiment 15.11 (p. 109)

D.

Discharge tube.

E.H.T. set.

Vacuum pump.

A very effective discharge tube can be made from a glass combustion tube as shown in Figure 15.6.

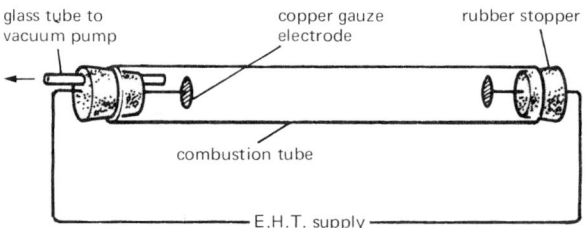

FIGURE 15.6

It is not intended that an exhaustive description of the various stages of the discharge should be given. It should be sufficient if the pupils observe the discharge in the partly evacuated tube and accept that it is accompanied by ultra-violet energy which, in turn, can make special powders lining discharge tubes fluoresce.

Pupils could be asked to find out from the school library the parts played by Swan and by Edison in the development of the incandescent filament lamp.

Electromagnetism

Experiments 15.12 – 20 (pp. 110–14)

P. 10 groups.

Westminster Electromagnetic kits.
High current output low voltage packs.

Pupils should already know the elementary facts about attraction and repulsion of magnetic poles and that (in their experience) only iron, steel, cobalt, and nickel are attracted by magnets, but it may be necessary to revise this work.

The power packs most suitable for pupil experiments are the special high current output types in which a resistance is incorporated in the output side so that they can be almost shorted without damage. If this type of power pack is not available lead-acid or nickel-iron accumulators can be used.

The experiments 15.12–15.16 on electromagnetism should be carried through at a fair pace and teachers should not make the going too slow by repeating the various experiments in too many guises. If the teacher

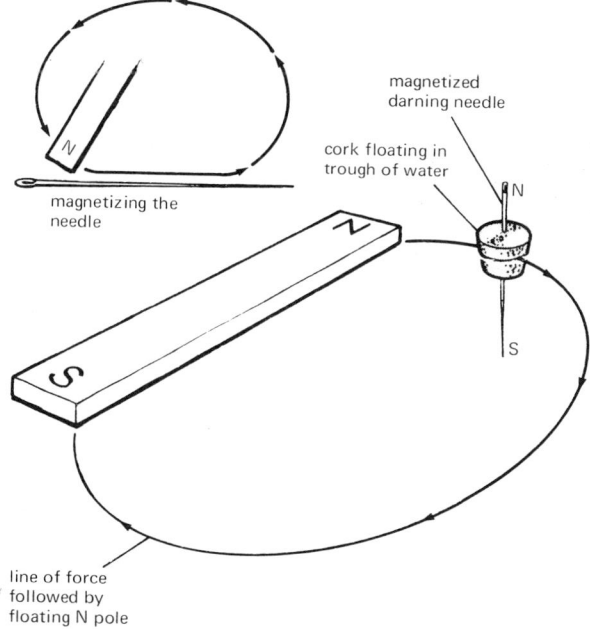

FIGURE 15.7

wishes to introduce additional experiments to those shown in the *Pupils' Textbook* it is suggested that every pupil should not be required to do each one unless there is plenty of time available.

The main point to bear in mind is that magnetism is here being approached as an effect of electrical energy and that knowledge of this effect has led to important discoveries in the production of motion (e.g. in the motor and the meter) and in the production of electrical energy on a large scale.

It is important that the pupils realize that the magnetic field of a coil exists only as long as current is flowing in the coil, unless a steel core is used when a permanent magnet would result if d.c. is used.

A very important feature to point out about lines of force is that they are like elastic threads along which the north-seeking pole pulls on the south-seeking pole, and that they never cross each other. Pupils should be encouraged to think of the lines being directional from N to S outside the magnet. This could be demonstrated by observing the path of a floating magnetized darning needle, Figure 15.7.

Force on a conductor
Experiments 15.17 – 15.18 (pp. 112 – 3)

In Figure 15.19 of the *Pupils' Textbook* the wire should be of fairly heavy gauge copper (16 s.w.g. approximately). It should, of course, be bare and have no kinks. The third piece should be tried as a roller and if it is to stay on the rails the two parallel wires must be level. Some teachers like to bend down 0·5 cm of each end of the third wire so that it does not fall off when once set in motion, but this usually requires a very high current output to work successfully.

Electric motor model
Experiment 15.19 (p. 113)

The successful winding of armatures as described in the *Pupils' Textbook* may be too difficult for some pupils. For these, pre-wound armatures could be made available so that, like the rest, they may achieve the satisfaction of successful accomplishment. Failing the use of the kit required in P.T. Figure 15.22, p. 114, the alternative model using cork and pins is easily constructed (UNESCO *Source Book for Science Teachers*, p. 168).

Pupils should be encouraged to bring old electric motors from toys such as trains, racing cars, etc., to school for examination.

The current meter

The pupils will have used ammeters in Unit 7. The type investigated here is the moving coil meter, which can be most easily recognized from its U-magnet, coil, and linear scale. To convert this type of meter from one range to another different shunts are connected across the coil in parallel to by-pass a proportion of the current. The length of wire used in the pupils' model should be long enough to provide about twenty turns on the wooden former. The 'springs' at each end should be wound in opposite directions. A milk straw provides the pointer.

The direction in which the coil turns will depend on the direction of the current and will be given by the left-hand rule.

A collection of moving-coil meters should be shown; the fuel gauge on a motor car dashboard is an ammeter which records the size of the current passing through a variable resistance the length of which is controlled by a float in the petrol tank. The lower the petrol level the greater the length of resistance wire in the circuit and so the lower the current.

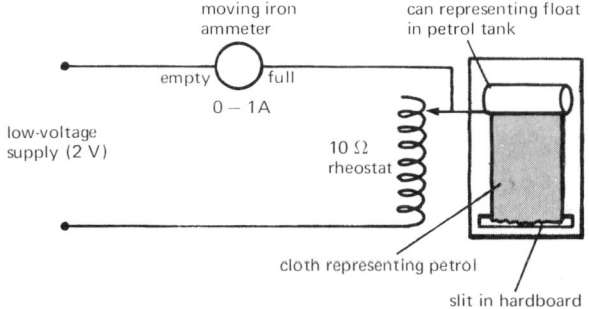

FIGURE 15.8 Model fuel gauge

The electric supply

Experiment 15.21 (p. 115)
Electrical energy from magnetism

P. 10 groups.

Apparatus shown in P.T. Figure 15.26 (p. 115).

Reference should be made to Faraday's experiments to try to show that electrical energy could be made from magnetism. Pupils could be asked to find out about the life of Faraday—an enthralling and encouraging story for any pupil—his upbringing as the son of a blacksmith, his minimal education, his first job as a message boy delivering books, his attendance at Davy's lectures, his job as a laboratory assistant, and his rise to be in charge of the Royal Institution Laboratory.

The meters used to detect the current induced in the coil in the pupils' experiments should have a range 1–0–1 milliamps.

Experiment 15.22 (p. 116)

P. 10 groups.

10 sets of apparatus shown in P.T. Figure 15.30 (p. 116).

Experiment 15.23 (p. 117)
The dynamo

P. 10 groups.

Motors made in earlier experiment.

10 galvanometers (or centre-zero meters).

Experiment 15.24 (p. 117)
The cycle dynamo

P. 10 groups if possible.

Cycle dynamos (as many as can be obtained).

Pupil type c.r.o.

Pupils might be asked to bring old cycle dynamos to school so that a collection of them, stripped down to their basic parts, can be examined. Manufacturers put on the market a cycle dynamo kit with which the pupils can make their own dynamos. In a cycle the return path from the lamps is usually through the cycle frame to the dynamo.

When a working cycle dynamo is connected to a c.r.o. with the time base switched off, the spot should move up and down as the coil rotates. If the dynamo is connected to a galvanometer and driven very slowly the needle should kick from side to side, indicating the change in direction of the current, and that it is producing alternating current. If the c.r.o. time base is switched on a jagged sine wave should be obtained from the dynamo output, again indicating a.c.

At this point it is useful to give the pupils the idea of the electrons moving under the influence of an alternating voltage. They reverse their direction of flow one hundred times every second, so that we have electrons flowing in one direction in the circuit for 0·01 second, and then in the opposite direction for 0.01 second. Each complete cycle, therefore, takes 0.02 second (or 1/50 second). We therefore have 50 cycles each second and the frequency is said to be 50 hertz.

Useful visual aids

The teacher will find it useful to see the Esso Film *The Westminster Electromagnetic Kit* in which the uses of the kit are shown. It is a 16-mm sound film obtainable from Travelling Films Ltd., 60–66 Wardour Street, London, W.1. For pupils, the film *Electromagnetic Induction* from Rank Film Library, 1 Aintree Road, Perivale, Greenford, Middlesex, gives a clear statement of the subject, and is useful for revision.

Useful *wall charts* are published by the Electrical Association for Women.

Appendix 1

Objective Test Items
Prepared by Dr. J. King
Principal Teacher of Chemistry, Wallace High School, Stirling in conjunction with the Scottish Education Department
and reproduced by kind permission of the Scottish Education Department and the Controller, H.M.S.O.

SECTIONS 9 – 15

An indication of the category of educational objective (*A*, *B*, *C*, or *D*) is given for each item (see *Curriculum Papers* 7, p. 27). The key response is indicated by a star*.

Choose the *best* answer in each case.

SECTION 9. Making Heat Flow

1. In an eight-storey block of flats, the boilers for central heating should be in

   *(A) the basement,   *B.*
   (B) the ground floor,
   (C) the fourth floor,
   (D) the top floor,
   (E) it does not matter.

2. Which of the following will be the best conductor?

   (A) Glass.   *A.*
   (B) Plastic.
   (C) Carbon.
   (D) Asbestos.
   *(E) Iron.

3. Eskimos can keep warm inside snow igloos because ice is a good

   (A) convector,   *C.*
   (B) conductor,

117

*(C) insulator,
(D) reflector of radiation,
(E) absorber of radiation.

4. A man builds a hut with an iron roof in a country with hot days and cold nights. The hut will be

(A) cool during the day,   *D.*
*(B) hot during the day,
(C) hot during the night,
(D) hot day and night,
(E) cool day and night.

5. A coal fire gives out heat mainly by

(A) conduction only,   *B.*
(B) convection only,
(C) radiation only,
(D) conduction and convection,
*(E) convection and radiation.

6. A man with unusually large feet finds that he cannot buy fur-lined boots in his size. Which would be the warmest of the following?

(A) Rubber boots.   *C.*
(B) Leather boots.
(C) Suede shoes with leather soles.
(D) Leather shoes with rubber soles.
*(E) Leather shoes with foam rubber lining.

7. Winds occur because air

*(A) which is cold is less dense than hot air   *B.*
(B) is always cold over the sea,
(C) is a better conductor than land,
(D) is a better conductor than sea,
(E) allows radiation to pass through it.

8. Heat loss by convection currents in a stoppered flask are prevented

(A) by silvering on the inside surface of the walls,   *A.*
(B) by silvering on the outer surface of the walls,
*(C) by the vacuum between the double walls,
(D) because the stopper is made of insulating material,
(E) by packing insulation between the walls and the case.

9. On a hot day in the sun it will be coolest to wear
   (A) a loose coloured shirt,   *C.*
   (B) a string vest,
   (C) a cotton vest,
   *(D) a loose white shirt,
   (E) nothing at all.

10. Which one of the following statements is true? Conduction
    (A) takes place best in gases,   *B.*
    (B) takes place best in liquids,
    (C) currents can be seen in liquids,
    *(D) cannot take place in a vacuum,
    (E) is fastest in a vacuum.

11. Which one of the following statements is true? Convection
    (A) only takes place in liquids,   *B.*
    (B) is faster than radiation,
    (C) is always slower than conduction,
    (D) takes place faster in a large volume of water than a small volume,
    *(E) is always faster in gases than in liquids.

12. Which one of the following statements is true? Radiation
    (A) travels by movement of particles,   *A.*
    (B) travels by movement of currents in a fluid,
    (C) can only travel in a vacuum,
    *(D) travels in straight lines,
    (E) can only travel in air.

13. A lemon meringue pie needs as hot an oven as possible. The best way to bake it would be
    *(A) on the top shelf,   *C.*
    (B) on the middle shelf,
    (C) on the bottom shelf,
    (D) with a thick metal baking tray underneath it,
    (E) with a thin metal baking tray underneath it.

14. Modern houses are often built with a double pane of glass in the windows because
    (A) glass is a good conductor of heat,   *C.*
    (B) air trapped between the panes prevents convection,

(C) radiation cannot pass through the trapped air,
(D) radiation passes better through the trapped air,
*(E) air trapped between the panes reduces conduction.

15. After a heavy snowfall, a boy sees that his house still has snow on the roof while the similar house next door has none. This is because

(A) his house has dull brown tiles,   *D.*
(B) the next door house has shiny green tiles,
*(C) his roof is better insulated than next door,
(D) his house is worse insulated than next door,
(E) his house has central heating.

16. It is possible to have both ice and boiling water in the same tube in the experiment illustrated because

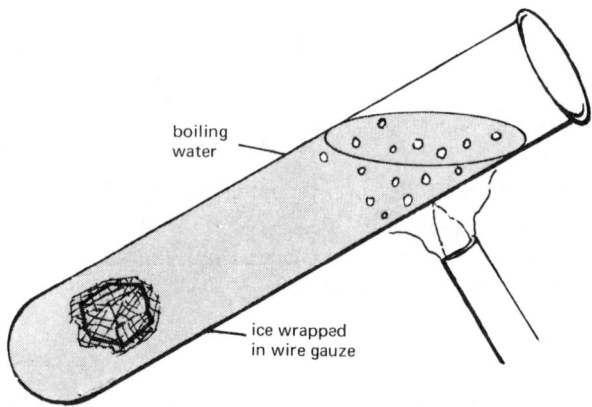

boiling
water

ice wrapped
in wire gauze

FIGURE T.1

(A) water is a good conductor,   *B.*
(B) glass is a good conductor,
*(C) hot water rises,
(D) the wire gauze insulates the ice,
(E) water is a poor convector.

17. When milk is boiled on a gas stove in a thin saucepan it sometimes burns. This does not happen with a thick saucepan because

(A) the heat cannot travel through a thick saucepan,   *C.*
(B) convection currents in the milk prevent burning,
*(C) the heat spreads out evenly before it reaches the milk,
(D) the thick saucepan is a better convector,
(E) the thick saucepan is a better radiator.

18. On a building site on a frosty morning, which will feel coldest to touch?

(A) Cement slabs.   *C.*
(B) Bricks.
(C) Wooden planks.
*(D) Iron scaffolding.
(E) Glass window panes.

19. Each of five similar tin lids is coated with different substances. Which will be the best heat absorber?

(A) White gloss paint.   *B.*
(B) Aluminium paint.
(C) Red paint.
(D) Black gloss paint.
*(E) Soot.

20. In question 19, which will be the best heat radiator?

(A) White gloss paint.   *B.*
(B) Aluminium paint.
(C) Red paint.
(D) Black gloss paint.
*(E) Soot.

21. Which one of the following statements is correct?

(A) Only conducted heat can pass through a vacuum.   *A.*
*(B) Only radiant heat can pass through a vacuum.
(C) Only convected heat can pass through a vacuum.
(D) Conducted and convected heat can pass through a vacuum.
(E) Convected and radiant heat can pass through a vacuum.

22. On a hot day, which car will become hottest inside?

(A) Shiny black.   *C.*
*(B) Dusty black.
(C) Shiny white.
(D) Dusty white.
(E) Red.

23. On a hot day, which car will be coolest inside?
    (A) Shiny black.   *C.*
    (B) Dusty black.
    *(C) Shiny white.
    (D) Dusty white.
    (E) Red.

24. One of the ways by which heat escapes from hot tea in a vacuum
    flask is radiation. Which of the following is most effective in
    preventing this loss?
    (A) Vacuum between the walls of the flask.   *A.*
    *(B) Silvering the flask walls.
    (C) Keeping the cork in the flask.
    (D) Keeping the flask in a metal case.
    (E) Insulating the flask from the metal case.

25.

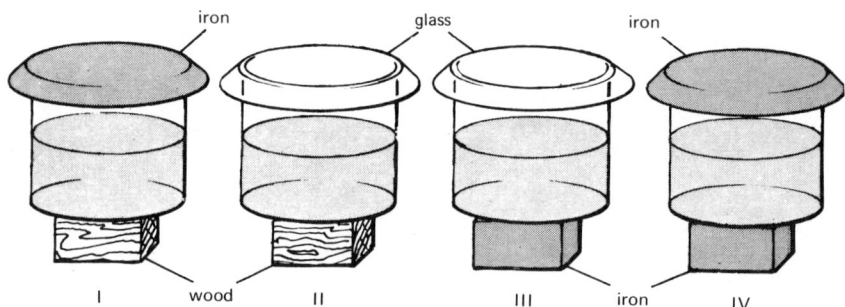

FIGURE T.2

Equal volumes of hot water are poured into each of the above
beakers. After five minutes, in which will the temperature have
dropped most?
    (A) All the same.   *B.*
    (B) I.
    (C) II.
    (D) III.
    *(E) IV.

26. Heat energy can be transferred
    X: by collision of particles,

Y: by currents in a fluid,
Z: through a vacuum.

Which of the following statements are true?

*(A) X, Y, and Z.   *A.*
 (B) Y and Z only.
 (C) X and Z only.
 (D) X and Y only.
 (E) Y only.

27.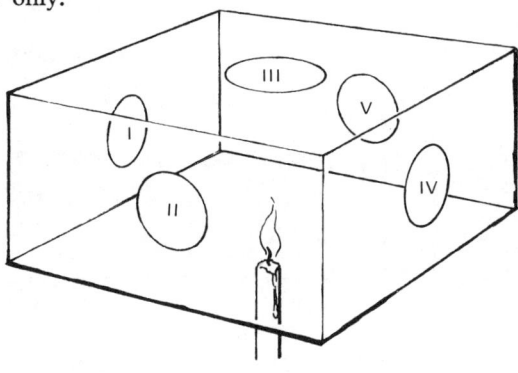

FIGURE T.3

In order for a candle to burn inside a box, in which two positions would it be best to punch holes?

 (A) I and II.   *C.*
*(B) II and III.
 (C) II and IV.
 (D) II and V.
 (E) I and V.

28.

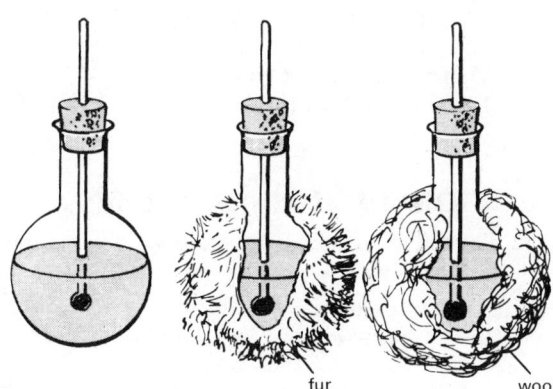

FIGURE T.4

In this experiment, boiling water is poured into three flasks. The temperature is then recorded every minute. The experiment

(A) shows that fur is a better insulator than wool,   *A.*
*(B) is bad because different volumes of water are used,
(C) shows that wool is a better insulator than fur,
(D) shows that fur and wool are both good insulators,
(E) shows that fur and wool are both useful for clothes.

29. Which diagram shows the convection currents correctly?   *A.*

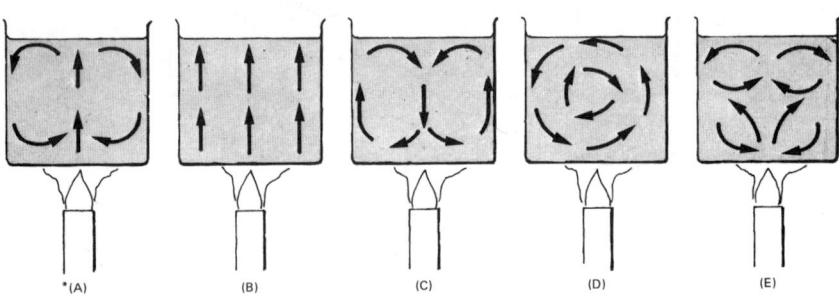

FIGURE T.5

30. Heat energy from a space craft in outer space can be lost by:

*(A) radiation,   *B.*
(B) conduction,
(C) convection,
(D) conduction and convection,
(E) radiation and convection.

SECTION 10. Hydrogen, Acids and Alkalis

31. The following metals are placed in a solution of dilute acid. Which will not react?

(A) Aluminium.   *A.*
*(B) Copper.

(C) Magnesium.
(D) Calcium.
(E) Iron.

32. The pH of a neutral solution is

   (A) 1,   *A.*
   (B) 5,
★(C) 7,
   (D) 9,
   (E) 10.

33. Which of the following will be shown to be acid when touched with dry pH paper?

   (A) Washing-up liquid.   *D.*
   (B) Solid citric acid.
★(C) Vinegar.
   (D) Sodium bicarbonate solution.
   (E) Milk.

34. A boy playing with a bottle from a chemistry set spills some acid on the carpet. The best remedy would be to treat it with

   (A) water,   *B.*
   (B) lemon juice,
   (C) vinegar,
★(D) sodium bicarbonate,
   (E) soap.

35. Water is a

   (A) mixture of hydrogen and oxygen,   *A.*
★(B) compound of hydrogen and oxygen,
   (C) mixture of hydrogen and nitrogen,
   (D) compound of nitrogen and oxygen,
   (E) compound of hydrogen, nitrogen, and oxygen.

36. Which of the following statements is true? Hydrogen

★(A) is lighter than air,   *A.*
   (B) is heavier than air,
   (C) is soluble in water,
   (D) supports combustion,
   (E) does not burn.

37. Balloons used for carrying passengers are not normally filled with hydrogen because

   (A) hydrogen is not light enough to lift a man off the ground,   C.
   (B) hydrogen is so light that the balloon would not come down again,
   (C) hydrogen has an unpleasant smell,
   *(D) the danger of explosion is too high,
   (E) the heat of the sun would burst the balloon.

38. A factory pumps out acid waste into a nearby river. Tests show that the pH of the river water is 8. This is because

   (A) the acid is very dilute when it is in the river,   D.
   (B) no acid has been pumped for some time,
   (C) the water is acidic,
   (D) the acid waste has been neutralized,
   *(E) nearby fields have been limed.

39. Air is a mixture of

   (A) hydrogen, oxygen, and carbon dioxide,   A.
   *(B) oxygen, nitrogen, and carbon dioxide,
   (C) hydrogen, nitrogen, and carbon dioxide,
   (D) hydrogen, oxygen, and nitrogen,
   (E) hydrogen, oxygen, and carbon dioxide.

40. 10 cm³ of dilute acid is placed in each of three beakers X, Y, and Z. 10 cm³ of dilute alkali is added to Y and 20 cm³ of dilute alkali is added to Z. Which will be the likely series of pH for X, Y, and Z in that order?

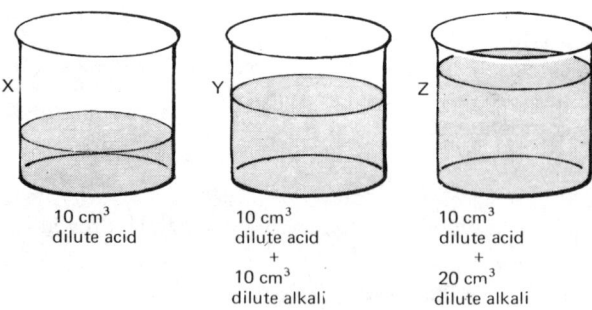

FIGURE T.6

(A) 7—9—5.   C.
(B) 9—7—5.
(C) 5—9—7.
(D) 9—5—7.
★(E) 5—7—9.

41. A test-tube rack has a number of labelled tubes containing solutions of known pH.

L	M	N	P	Q	R
3	4	5	7	8	9

P could be made more alkaline by

(A) dilution,   D.
(B) addition of L,
(C) addition of N,
★(D) addition of Q,
(E) evaporation.

42. Given the following solutions

L	M	N	P	Q	R
3	4	5	7	8	9

with which of the following mixtures would it be possible to obtain a neutral solution?

(A) L+M.   C.
(B) L+P.
★(C) L+R.
(D) P+Q.
(E) Q+R.

43. A chemist wishes to arrange five solutions in order of their acidity. He has a universal indicator chart

red   4 or less
orange 5
yellow 6
green   7
blue   8
indigo 9
violet 10

and obtains the following results: L orange/yellow
                                    M green

N green/blue
P green/yellow
Q blue

Which is the correct order for decreasing acidity?

(A) L M N P Q.
(B) L N M P Q.
(C) L P M Q N.
(D) L P N M Q.
*(E) L P M N Q.

D.

44. Some metals can be separated from a mixture of them with gold, by dissolving them in dilute hydrochloric acid. For which of the following would this NOT be possible?

(A) Tin.
(B) Iron.
(C) Zinc.
*(D) Copper
(E) Magnesium.

C.

45. Which of the following will react most quickly when placed in water?

*(A) Sodium.
(B) Magnesium ribbon.
(C) Magnesium powder.
(D) Aluminium.
(E) Calcium.

A.

46. 10 cm³ of a given alkali solution is placed in each of four beakers W, X, Y, and Z.

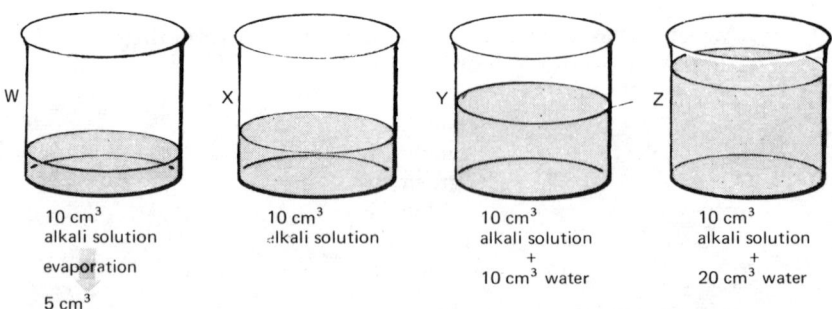

W
10 cm³
alkali solution
evaporation
5 cm³
alkali solution

X
10 cm³
alkali solution

Y
10 cm³
alkali solution
+
10 cm³ water

Z
10 cm³
alkali solution
+
20 cm³ water

FIGURE T.7

W is evaporated to half its volume.
X is unchanged.
Y has 10 cm³ of water added.
Z has 20 cm³ of water added.

Which beaker will require most of a given acid solution for neutralization?

(A) W.   *D.*
(B) X.
(C) Y.
(D) Z.
*(E) All the same.

47. 50 cm³ of dilute hydrochloric acid is placed in each of three beakers X, Y, and Z. A piece of calcium is added to X, a piece of zinc to Y, and some mercury is added to Z.

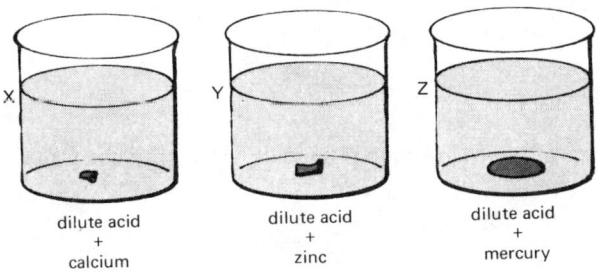

FIGURE T.8

Which beaker will have the most acid remaining?

(A) X.   *D.*
(B) Y.
*(C) Z.
(D) X and Y.
(E) Y and Z.

48. The pH of soil in which various plants prefer to grow can be measured. In one study, the following results were obtained.

Plant	pH preferred
Wheat	6.0–7.5
Rye	4.8–5.2
Oats	4.8–6.3
Swedes	4.7–5.2
Barley	6.7–7.5

Which of the following would be likely to grow well together in the same alkaline field?

(A) Oats and swedes.   *D.*
(B) Rye and oats.
(C) Wheat and oats.
(D) Oats and barley.
★(E) Wheat and barley.

49. During a chemical reaction, a gas was given off which turned lime water milky. This showed that the gas was

(A) nitrogen,   *A.*
(B) water vapour,
★(C) carbon dioxide,
(D) oxygen,
(E) hydrogen.

50. Which of the following will burn in air to produce water?

(A) Nitrogen.   *A.*
(B) Water vapour.
(C) Oxygen.
★(D) Hydrogen.
(E) Carbon dioxide.

51. The best way to identify a liquid as being water is to

(A) taste it,   *A.*
(B) add calcium,
(C) test it with pH paper,
★(D) show that it boils at 100° C,
(E) show that it can put out fires.

52. Sodium is stored

(A) in a bottle under water,   *C.*
★(B) in a bottle under oil,
(C) lying in an open box,

(D) in a gas cylinder,
(E) in a dropping bottle.

53. The soil in a garden was found to be very acidic. To cure this the gardener would add

(A) water, *A.*
(B) salt,
(C) fertilizer,
*(D) lime,
(E) insecticide.

54. A metal reacts slowly with dilute acid but not with water. The metal is likely to be

(A) calcium, *A.*
(B) sodium,
(C) mercury,
*(D) zinc,
(E) gold.

55. Which of the following would be most suitable for making water pipes?

(A) Calcium. *C.*
*(B) Aluminium.
(C) Magnesium.
(D) Mercury.
(E) Sodium.

56. Which of the following best explains what happens when sodium is added to water?

(A) The sodium melts in the water. *D.*
(B) The sodium dissolves in the water.
*(C) The sodium reacts with the water.
(D) The water absorbs the sodium.
(E) The sodium floats on the water.

57. The following facts are known about three metals X, Y, and Z.

Z tarnishes quickly in air while X and Y do not.

Z and X react with dilute hydrochloric acid to give hydrogen while Y does not.

The order of activity, most reactive first, is

*(A) Z X Y. *C.*
(B) X Y Z.

(C) Y Z X.
(D) Z Y X.
(E) Y X Z.

58. A wasp sting is alkaline and so could be countered by dabbing with a solution of:

(A) soap,   *C.*
(B) salt,
(C) washing soda,
★(D) vinegar,
(E) ammonia.

59. Acid has been spilled on someone's clothes. The best treatment would be to:

(A) wash them with lots of water,   *B.*
(B) wash them with vinegar,
(C) send them to the cleaners,
★(D) wash them with sodium bicarbonate solution,
(E) do nothing until it dries.

60. These lists of metals are supposed to be in the order of their decreasing vigour of reaction with water. Which one is correct?

(A) calcium—sodium—iron—magnesium.   *A.*
(B) sodium—magnesium—iron—calcium.
(C) magnesium—iron—sodium—calcium.
(D) iron—sodium—calcium—magnesium.
★(E) sodium—calcium—magnesium—iron.

61. All three metals in one of these lists displace hydrogen from dilute acids. Which list is correct?

(A) lead—magnesium—aluminium.   *A.*
★(B) magnesium—aluminium—iron.
(C) copper—aluminium—iron.
(D) silver—magnesium—aluminium.
(E) mercury—silver—copper.

62. A gas lights with a pop and burns with a blue flame. It is likely to be

(A) oxygen,   *A.*
★(B) hydrogen,
(C) carbon dioxide,
(D) air,
(E) nitrogen.

SECTION 11. Detecting the Environment

63. When light enters the eye and falls on the retina it produces chemical changes in the cells of the retina. The result of this is that

    (A) the pupil of the eye gets larger,   *A.*
    ★(B) electrical signals are sent to the brain,
    (C) the lens increases in curvature,
    (D) the lens decreases in curvature,
    (E) the iris changes colour.

64. An eye looks at the image of a burning candle in a pinhole camera. The candle flame will point

    (A) up in the camera and down on the retina,   *D.*
    (B) up in the camera and up on the retina,
    ★(C) down in the camera and up on the retina,
    (D) down in the camera and down on the retina,
    (E) it depends on the eye.

65. The part of the eye which focuses light is the

    (A) retina,   *A.*
    (B) iris,
    ★(C) lens,
    (D) pupil,
    (E) optic nerve.

66. Which of the following represent correctly the path of a ray of light through a lens?

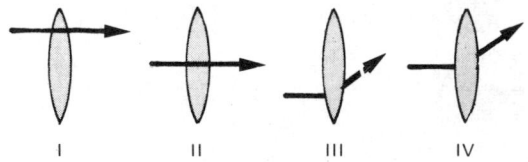

I   II   III   IV

FIGURE T.9

    (A) I and II.   *A.*
    (B) I and III.
    (C) I and IV.
    ★(D) II and III.
    (E) III and IV.

67. Which of the following statements is true? In a pinhole camera

   (A) only a large hole gives a sharp image,   *A.*

   ★(B) only a small hole gives a sharp image,

   (C) only a large number of small holes give a sharp image,

   (D) only near objects can be focused,

   (E) only distant objects can be focused.

68. A lens camera is to be used to photograph a galloping horse on a dull day. Which of the following will give the clearest picture?

   (A) Long exposure, wide aperture.   *D.*

   ★(B) Short exposure, wide aperture.

   (C) Long exposure, small aperture.

   (D) Short exposure, small aperture.

   (E) Medium exposure, medium aperture.

69. Which of the following is true? The amount of light entering the eye is regulated by the

   ★(A) iris,   *A.*

   (B) ciliary muscles,

   (C) lens,

   (D) cornea,

   (E) retina.

70. Which of the following are examples of optical illusions?

   X: movement in a motion picture.

   Y: the colours produced from white light through a prism.

   Z: darkness as one goes indoors from bright sunshine.

   (A) X only.   *A.*

   (B) Y only.

   (C) Z only.

   (D) X and Y.

   ★(E) X and Z.

71. A cricketer loses the sight of one eye in a car crash. The other is unaffected. When he plays cricket, the effect will be that

   ★(A) he cannot judge how far away the ball is,   *B.*

   (B) he can only see the ball half the time,

   (C) he cannot focus on the ball,

   (D) the ball looks smaller,

   (E) he cannot follow the flight of the ball at speed.

72. Tom is colour-blind. This means that

  (A) he cannot see coloured objects,   *A.*
  (B) he sees everything in black and white,
  (C) all colours look the same,
 *(D) he confuses some colours only,
  (E) everything looks grey.

73.

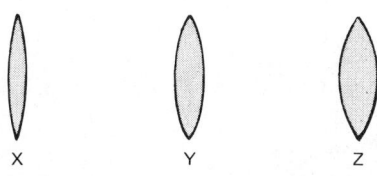

FIGURE T.10

These represent the shape of the lens of a woman's eye. She holds a needle close to her eye to thread it, then looks at the sewing on her lap, then at the TV across the room. Which is the correct sequence for the shape of the lens during these activities?

  (A) X Y Z.   *D.*
  (B) X Z Y.
  (C) Y X Z.
  (D) Z X Y.
 *(E) Z Y X.

74. A boy walks through a green field looking for some flowers. He finds none until he bends down to tie his shoelace and discovers that there are hundreds all around him. He picks a bunch of blue ones but misses the similar red ones growing around them. This shows that he is

  (A) long sighted,   *D.*
  (B) short sighted,
  (C) colour blind,
  (D) long sighted and colour blind,
 *(E) short sighted and colour blind.

75. The stirrup, anvil, and hammer bones are found in the

  (A) outer ear,   *A.*
 *(B) middle ear,

(C) inner ear,
(D) cochlea,
(E) eardrum.

76. A musician wishes to change the pitch of a violin string. He could do this by

(A) changing the tension in the string,   *C.*
(B) lengthening the vibrating distance,
(C) shortening the vibrating distance,
(D) altering the frequency of vibration,
*(E) any of these.

77. A ringing electric bell is placed in a bell jar and the air is pumped out. The sound will

(A) increase in loudness,   *A.*
*(B) decrease in loudness,
(C) remain the same loudness and pitch,
(D) decrease in pitch,
(E) increase in pitch.

78. In the experiment in question 77, the evacuated bell jar is slowly filled with water. The sound will then

*(A) increase in loudness,   *A.*
(B) decrease in loudness,
(C) remain the same loudness and pitch,
(D) decrease in pitch,
(E) increase in pitch.

79. If a drum-skin is tightened, its note will

*(A) increase in pitch,   *B.*
(B) decrease in pitch,
(C) sound louder,
(D) sound quieter,
(E) stay the same pitch and loudness.

80. From the given graph for human hearing in Figure T.11, which of the following deductions could one make?

(A) High frequency sounds are easier to hear than low frequency sounds.   *C.*
(B) Low frequency sounds are easier to hear than high frequency sounds.

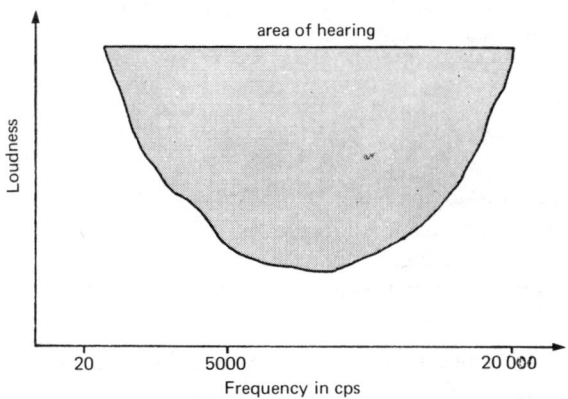

FIGURE T.11

*(C) The threshold frequencies of sound are 20 and 20 000 cps.
(D) Frequencies below 20 cps can only be heard if they are very loud.
(E) Frequencies of 5000 cps can be heard no matter how quiet they are.

81. Between which of the following pairs would it be possible to distinguish by taste alone?
(A) Lemon juice and vinegar. B.
(B) Honey and treacle.
*(C) Barley sugar and acid drops.
(D) Strawberry and vanilla ice cream.
(E) Black coffee and black tea.

82. The four fundamentally different tastes are
*(A) sweet, sour, salt, bitter, A.
(B) sweet, sour, creamy, salt,
(C) sweet, bitter, creamy, salt,
(D) sweet, bitter, meaty, salt,
(E) sweet, sour, meaty, creamy.

83. A man with a heavy cold finds his food tasteless because
(A) the cold germs kill the taste of the food, B.
(B) the cold germs numb the taste buds,
*(C) his blocked nose cuts out his sense of smell,

(D) the nerve endings on his tongue are affected,

(E) he has lost his appetite.

84. A man can wear glasses without irritating his nose because the bridge of the nose

    (A) has no nerve endings,   *C* (*A?*)

    (B) has no pressure spots,

  ★(C) becomes accustomed to pressure,

    (D) can only distinguish between hot and cold,

    (E) can only feel pain and not just light pressure.

85. Three men enter a room whose temperature is 25° C.

    X has been sitting in a room at 20° C.

    Y has been out in the garden at 5° C wearing a thick sweater and an anorak.

    Z has been out in the garden at 5° C wearing a thin sweater.

    To which, if any, will the room feel hottest?

    (A) X.   *D.*

    (B) Y.

  ★(C) Z.

    (D) Y and Z.

    (E) X, Y, and Z.

86. A boy is told to shut his eyes and some food is put on his tongue. He would best be able to guess what it was by using

    (A) taste only,   *A.*

    (B) smell only,

    (C) taste and feel,

    (D) taste and smell,

  ★(E) taste, smell, and feel.

87. A girl who is going out to a dance sprays some long-lasting perfume behind her ears. A few minutes later she realises she cannot smell it any more. This is because

    (A) it has all evaporated,   *C.*

    (B) she cannot smell behind her ears,

    (C) perfume only smells for a very short time,

  ★(D) her sense of smell has been deadened to that smell,

    (E) her sense of smell is very poor.

88. The crews of ships at sea can communicate by shouting through loudhailers. It is impossible for the crews of spaceships in space to do this because

    (A) the temperature is too low,   *C.*
    (B) the sound is reflected,
    (C) the pressure is too high inside the spaceship,
    (D) the sound barrier has been broken,
    *(E) there is no air.

89. A patient is told by his doctor to allow a pill to dissolve slowly in his mouth. If the pill is unpleasantly bitter, the best way to take it would be to

    (A) put it at the front of the tongue,   *C.*
    (B) put it in the middle of the tongue,
    *(C) put it in the cheek,
    (D) hold the nose while taking it,
    (E) it would not matter.

90. A scientist believes that cats may not like the cold. To test his hypothesis he should study what happens when he

    (A) puts a cat in a warm room,   *C.*
    (B) puts a cat in a cold room,
    (C) lets a cat choose between hot and cold rooms,
    *(D) lets several cats choose between hot and cold rooms,
    (E) puts several cats in a cold room.

91. Bones are present in the middle ear in order to

    (A) stop the eardrum from collapsing,   *B.*
    *(B) transmit sound vibrations,
    (C) stop the ear from vibrating too much,
    (D) protect the inner ear from dirt,
    (E) transmit the feeling of balance.

92. Tea tasters are people who can distinguish the fine differences between different grades of tea. They sip each blend of tea and then spit out the mouthful. The best taster would be

    (A) someone who was always thirsty,   *C.*
    (B) someone whose favourite drink was tea,
    *(C) someone whose taste buds were very sensitive,
    (D) someone with a poor sense of smell,
    (E) a simple instrument for measuring taste.

SECTION 12. The Earth

93. Which of the following does NOT contain the element silicon?

    (A) Clay.   *A.*

    (B) Sand.

    (C) Quartz.

    (D) Glass.

   ★(E) Malachite.

94. Which of the following metals are both found uncombined in nature?

    (A) Magnesium; zinc.   *A.*

    (B) Magnesium; silver.

   ★(C) Copper; silver.

    (D) Copper; magnesium.

    (E) Zinc; silver.

95. A great deal of energy is released when a certain metal is burned in oxygen. This suggests that the

    (A) metal oxide could be easily decomposed by heating,   *D.*

    (B) metal will be found uncombined in nature,

   ★(C) metal oxide would need a lot of energy to free the metal,

    (D) .metal will be obtained from the oxide by heating with carbon,

    (E) metal must be magnesium.

96. When iron pyrites (iron sulphide) is heated in air

    (A) there is no change,   *A.*

    (B) sulphur and iron are produced,

    (C) sulphur dioxide and iron are produced,

    (D) sulphur and iron oxide are produced,

   ★(E) sulphur dioxide and iron oxide are produced.

97. After evaporating sea water, a pupil performs a flame test on the residue and finds that the flame is yellow. This suggests that sea water contains

    (A) sodium chloride,   *D.*

   ★(B) sodium salts,

    (C) potassium chloride,

    (D) potassium salts,

    (E) sodium and potassium salts.

98. Rocks which have been formed, then changed in their nature by the effects of great pressure and heat, are called

   (A) igneous,   *A.*
 *(B) metamorphic,
   (C) sedimentary,
   (D) carboniferous,
   (E) fossilized.

99. Marble is a naturally occurring form of

   (A) copper carbonate,   *A.*
   (B) magnesium oxide,
 *(C) calcium carbonate,
   (D) copper oxide,
   (E) calcium oxide.

100. The Bronze Age (bronze is an alloy of copper and tin) came before the Iron Age in history. Copper was probably discovered before iron because

   (A) copper compounds are more plentiful in the earth's crust,   *C.*
 *(B) copper compounds need less energy to release the copper,
   (C) iron compounds need less energy to release the iron,
   (D) copper compounds occur nearer the earth's surface,
   (E) iron ores are not affected by charcoal (carbon).

101. Crude oil can be separated by fractional distillation because each fraction has a different

   (A) composition,   *B.*
   (B) density,
 *(C) boiling point,
   (D) melting point,
   (E) viscosity.

102. Some soil is shaken up with water in a large jar. The order in which the soil particles settle out is

 *(A) gravel; coarse sand; fine sand; clay,   *A.*
   (B) clay; gravel; fine sand; coarse sand,
   (C) gravel; clay; fine sand; coarse sand,
   (D) clay; coarse sand; fine sand; gravel,
   (E) coarse sand; fine sand; clay; gravel.

103. A piece of Roman pottery dug up in a ruin has a blue-green glaze.

This might suggest that Roman potters used glazes containing compounds of

(A) sodium,   *C.*
(B) zinc,
(C) calcium,
(D) lead,
*(E) copper.

104. Heating coal in a limited supply of air gives a material commonly used in industry for obtaining metals from their ores. This material is

(A) tar,   *B.*
(B) ammonia,
(C) coal gas,
*(D) coke,
(E) sulphur.

105. An element which can exist uncombined in the earth's crust is

(A) potassium,   *A.*
(B) phosphorus,
(C) sodium,
(D) magnesium,
*(E) sulphur.

106. A planet has an atmosphere consisting of nitrogen only. Which of the following uncombined elements might be found pure in its crust?

(A) Iron.   *C/D.*
(B) Lead.
(C) Tin.
*(D) All of these.
(E) None of these.

107. In an article in a newspaper on central heating, mention was made of 'fossil' fuel. This was probably referring to

(A) hydroelectric power,   *C.*
(B) direct sunlight,
(C) wood and paper,
*(D) coal and oil,
(E) atomic energy.

108. Heating coal out of contact of air gives rise to a gas called

    (A) natural gas,   *B.*
  ★(B) coal gas,
    (C) coke gas,
    (D) calor gas,
    (E) none of those.

109. A night watchman notices that a certain type of coal on his fire gives a sharp, acid smell. This is probably due to the formation of

  ★(A) sulphur dioxide,   *C.*
    (B) nitrogen oxide,
    (C) carbon monoxide,
    (D) carbon dioxide,
    (E) ammonia.

110. Milk containing micro-organisms can be purified by

    (A) filtering,   *A.*
    (B) freezing,
    (C) cooling,
    (D) warming,
  ★(E) boiling.

111. Which of the following statements is NOT true?

    (A) Cheese results from the growth of micro organisms in sour milk.   *A.*
    (B) Different kinds of cheese can be made using different types of micro-organism.
  ★(C) In cheese production, micro-organisms are only needed to give flavour.
    (D) In butter production, micro-organisms are only needed to give flavour.
    (E) Unleavened (unrisen) bread is made without yeast.

112.

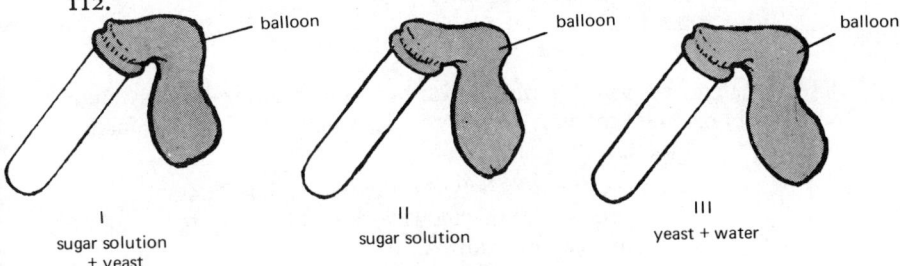

         I   II   III
sugar solution   sugar solution   yeast + water
  + yeast

FIGURE T.12

The three test-tubes in Figure T.12 were placed in a water bath at 37° C. After 30 minutes a gas had filled balloons

    (A) I, II, and III,   *B.*
    (B) I and II,
    (C) I and III,
 *(D) I only,
    (E) III only.

113. Micro-organisms are NOT required in the manufacture of

    (A) beer,   *A.*
    (B) cheese,
    (C) bread,
 *(D) jam,
    (E) yoghurt.

114. Which of the following statements is correct?
Worms help to improve the soil because they

    X: increase the drainage.
    Y: churn the soil up.
    Z: digest vegetable material.

 *(A) X, Y, and Z.   *A.*
    (B) X and Y.
    (C) X and Z.
    (D) Y only.
    (E) Z only.

115. Addition of sulphuric acid to malachite results in

    (A) copper oxide+carbon dioxide,   *A/B.*
    (B) copper oxide−sulphur dioxide,
    (C) copper sulphate+sulphur dioxide,
 *(D) copper sulphate+carbon dioxide,
    (E) copper+sulphur dioxide.

116. A scientist is asked to distinguish between a sample of magnesium carbonate and calcium carbonate. He would do this by means of

 *(A) a flame test,   *B.*
    (B) the colour difference between the two carbonates,
    (C) reducing them to their metals with carbon,
    (D) adding dilute hydrochloric acid,
    (E) heating them.

117. Analysis of a sample of water showed that it contained salts of sodium and calcium. It would be likely to have come from

  *(A) the sea,   *C.*
   (B) a reservoir in a limestone region,
   (C) a reservoir in a chalk region,
   (D) a river flowing through a region of salt deposits,
   (E) a river flowing through a region of peat bogs.

118. A ship sailing round the coast of Iceland discovers a newly-formed volcanic island a few miles from the mainland. The rocks of which the island is composed will be mainly

   (A) metamorphic,   *C.*
  *(B) igneous,
   (C) sedimentary,
   (D) metamorphic and igneous,
   (E) metamorphic and sedimentary.

119. In New Zealand, hot water springs flowing over rocks composed of silicates

   (A) dissolve them completely,   *C.*
   (B) change them to silica,
   (C) change them to silicon,
  *(D) hardly affect them at all,
   (E) make them become much softer.

120.

Oxide of metal	Effect of heat on oxide	Effect of carbon and heat on oxide
X	decomposed	decomposed
Y	nil	nil
Z	nil	decomposed

From the above table, the order of activity of the three metals is

   (A) X Y Z,   *D.*
   (B) X Z Y,
   (C) Y Z X,
  *(D) Y X Z,
   (E) Z Y X.

121. The two most common elements in the earth's crust are
    *(A) oxygen and silicon,   A.
     (B) oxygen and aluminium,
     (C) silicon and aluminium,
     (D) iron and silicon,
     (E) oxygen and iron.

122. The energy released when peat is burned came *originally* from
    *(A) the sun,   C.
     (B) plants,
     (C) the soil,
     (D) the atmosphere,
     (E) the sea.

SECTION 13. Support and Movement

123. Which of the following are examples of forces acting?
       X: A footballer kicking a ball.
       Y: A goalkeeper stopping a goal.
       Z: A referee blowing his whistle.

     (A) X only.   B.
     (B) X and Y only.
     (C) X and Z only.
     (D) Y and Z only.
    *(E) X, Y, and Z.

124. Which of the following are examples of forces acting?
       X: the attraction of two electrically charged plastic rods.
       Y: clothes blowing on a clothes line.
       Z: a man leaving footprints in the snow.

     (A) X only.   B.
     (B) X and Y only.
     (C) X and Z only.
     (D) Y and Z only.
    *(E) X, Y, and Z.

125. An object requires force to
     (A) start it moving,   A.
     (B) change its direction of movement,
     (C) change its shape,

(D) stop it from moving,
★(E) bring about all of these.

126. Which of the following make use of friction?

(A) Brakes on a car.   *B.*
(B) A parachute.
(C) Grit on icy roads.
(D) Oars on a rowing boat.
★(E) All of these.

127.

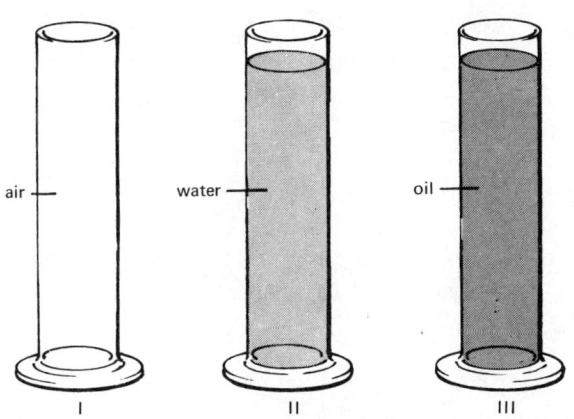

FIGURE T.13

A steel ball bearing takes: 1 second to fall in I.
2 seconds to fall in II.
20 seconds to fall in III.

This experiment shows that:

(A) the force of gravity is greatest in I.   *C.*
(B) the force of gravity is greatest in III.
(C) the force of friction is greatest in I.
★(D) the force of friction is greatest in III.
(E) friction only occurs in liquids.

128. When a sledge is pulled up a hill as in Figure T.14, the direction
of the force of friction is shown by arrow

(A) X.   *B.*

  *(B) Y.
   (C) Z.
   (D) W.
   (E) None of these.

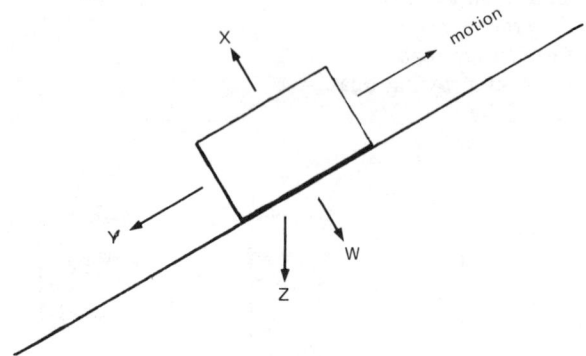

FIGURE T.14

129. A body moves in one direction at constant speed. If no force is
     applied the body will

     (A) speed up,   *B.*
     (B) slow down,
   *(C) continue at the same speed,
     (D) change its direction,
     (E) change its shape.

130. A stone thrown into the air returns to earth because

   *(A) it is attracted by the mass of the earth,   *A.*
     (B) the atmosphere presses it back down,
     (C) the earth's spin sucks it back,
     (D) stones do not contain any air,
     (E) it is attracted by the earth's magnetism.

131.

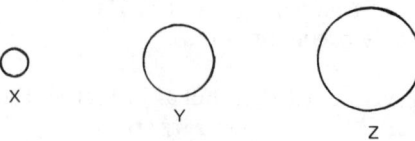

FIGURE T.15

The three planets X, Y and Z are made of the same material. A block of lead is weighed on each planet in turn. The order of weight (lightest first) is

*(A) X, Y, Z,   D.
(B) X, Z, Y,
(C) Z, X, Y,
(D) Z, Y, X,
(E) all the same.

132. Which of the following is the unit for measuring force?

(A) Pound.   A.
(B) Joule.
*(C) Newton.
(D) Kilogramme.
(E) Gramme.

133.

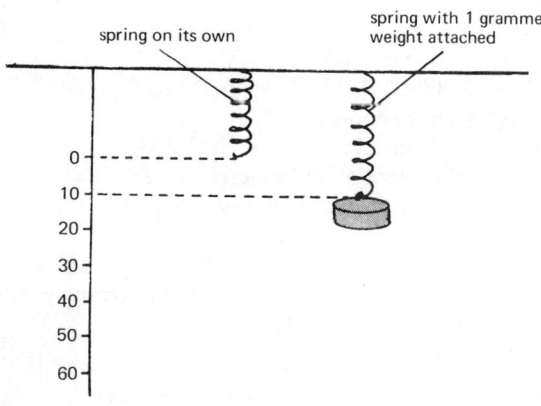

spring on its own

spring with 1 gramme weight attached

0
10
20
30
40
50
60

FIGURE T.16

When 5 grammes are placed on the spring instead of 1 gramme, the reading will be

(A) 15,   B.
(B) 25,
(C) 40,
*(D) 50,
(E) 60.

134. Which of the following show that forces occur in pairs?

    (A) The recoil of a gun.   *A.*
    (B) Water thrown back as a water rocket takes off.
    (C) Two trucks move apart on a track after colliding.
  *(D) All of these.
    (E) None of these.

135. 1 joule (J) equals

    (A) 1 newton,   *A.*
    (B) 1 newton force,
    (C) 1 newton centimetre,
  *(D) 1 newton metre,
    (E) 1 newton/metre.

136. How much work is done when a load of 12 N is raised 2 metres?

  *(A) 24 N m.   *B·*
    (B) 24 N per m.
    (C) 14 N.
    (D) 6 N m.
    (E) 6 N per m.

137. A car moving uphill at a steady speed is

  *(A) gaining potential energy,   *C.*
    (B) gaining kinetic energy,
    (C) gaining both potential and kinetic energy,
    (D) neither gaining nor losing energy,
    (E) losing energy.

138. A man weighing 1000 N possesses 8000 J of potential energy. How high is the man above ground level?

    (A) 8000 m.   *D.*
    (B) 800 m.
    (C) 80 m.
  *(D) 8 m.
    (E) 0 m.

139. A ball is dropped from a window 11.5 m above the ground. At which level will the potential energy be slightly greater than the kinetic energy?

  *(A) 6m.   *D.*

(B) 5m.
(C) 4m.
(D) 2m.
(E) 0m.

140. A man was trying to remove a nut using a spanner but found he could not exert enough force. The best thing to do would be to use a

(A) lighter spanner,   *B.*
(B) shorter spanner,
(C) heavier spanner,
(D) spanner with a larger head,
★(E) longer spanner.

141. If a body is moving in a straight line with a constant speed, which of the following statements MUST be FALSE?

(A) There are no forces acting on the body.   *B.*
(B) There may be forces acting on the body but they cancel each other out.
★(C) There can only be a force acting at right angles to its path.
(D) The body has no acceleration or deceleration.
(E) There is no frictional force.

142. The frictional force acting on a block of wood is known to be 10 N. The force required to pull the block along the bench at a steady speed is

(A) 11 N,   *B.*
★(B) 10 N,
(C) 9 N,
(D) 5 N,
(E) 0 N.

143.

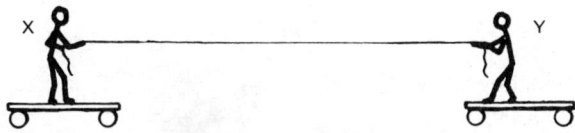

FIGURE T.17

Two pupils are standing on trolleys and each is holding the end of a rope. When pupil X pulls the rope

(A) only pupil X moves,　　　　　　　　　　　　　*A.*
(B) only pupil Y moves,
★(C) both move together,
(D) pupil X moves forward and pupil Y moves backwards,
(E) pupil Y moves forwards and pupil X moves backwards.

144. Thirty passengers are travelling on a double decker bus along a motorway in a high cross-wind. Which will be the most stable arrangement of the passengers?

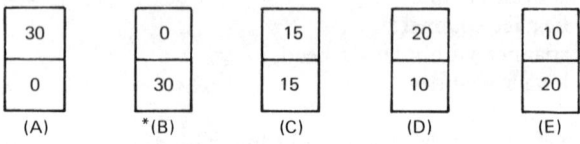

FIGURE T.18

145. The work done when a force F is applied through a distance d is

(A) $\dfrac{F}{d}$,　　　　　　　　　　　　　　　　　　*A.*

(B) $\dfrac{d}{F}$,

(C) $F+d$,
(D) $F-d$,
★(E) $F \times d$.

146.

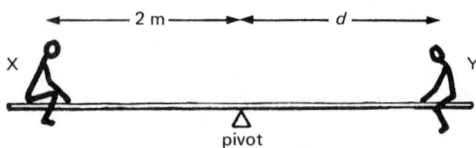

FIGURE T.19

A 50 kg boy sits at X.
A 40 kg boy sits at Y.
To keep the see-saw balanced, distance d must be

(A) 1.6 m,　　　　　　　　　　　　　　　　　　*B.*
★(B) 2.5 m,
(C) 5 m,

(D) 20 m,
(E) none of these.

147. Which of the following animals has no backbone?
    (A) Fish.   *A.*
    (B) Man.
    ★(C) Beetle.
    (D) Seal.
    (E) Snake.

148. Which of the following chairs is the most stable?

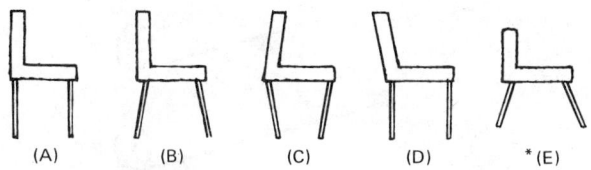

FIGURE T.20

149. Which of the following possesses an external-skeleton?
    (A) Earthworm.   *A.*
    ★(B) Locust.
    (C) Whale.
    (D) Dog.
    (E) Sparrow.

150.

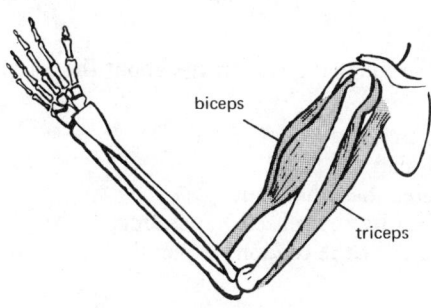

FIGURE T.21

To straighten out the arm, the
(A) biceps must contract,   *A.*

*(B) triceps must contract,
(C) biceps must expand,
(D) biceps and triceps must contract,
(E) biceps and triceps must expand.

151. In which of the following situations is it easiest to move the
     rock R:

D.

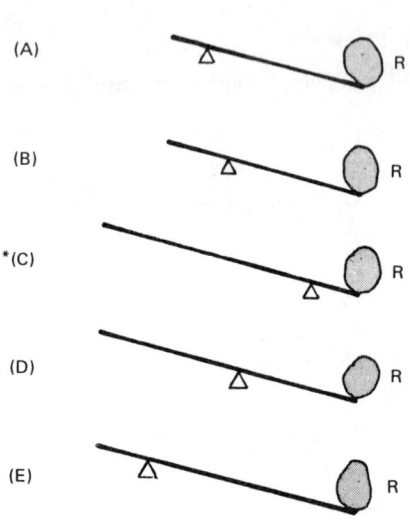

FIGURE T.22

152. Which of the following statements about the force in the biceps is
     correct? It

(A) equals the load,   *B.*
(B) is less than the load,
*(C) is greater than the load,
(D) is caused by the muscle expanding,
(E) is caused by the tendons expanding.

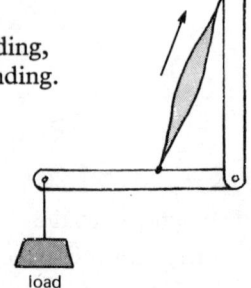

FIGURE T.23   load

153.

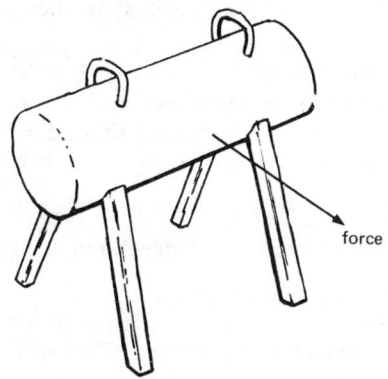

FIGURE T.24

The amount of force needed to knock over a horse in a gym is found for various arrangements of the legs, as shown.

		Legs long	Legs short
Legs outward		700 units	800 units
Legs straight		600 units	700 units
Legs inward		400 units	500 units

The most stable arrangement is

(A) outwards, long,   D.
(B) straight, long,
(C) inwards, short,
⋆(D) outwards, short,
(E) straight, short.

154. Given the data in question 153, which will be the least stable arrangement of the legs?

(A) Outwards, long.   D.
(B) Straight, long.
⋆(C) Inwards, long.
(D) Outwards, short.
(E) Inwards, short.

155. Which of the following does NOT show the existence of gravity?

    (A) A meteorite speeds up when approaching the earth.   *B.*
  ★(B) A meteorite burns up when entering the earth's atmosphere.
    (C) The earth moves round the sun.
    (D) A stone thrown up in the air slows down as it rises.
    (E) An object weighs less the further it is moved away from the earth.

156. Which of the following are true? A moving rocket

    X. needs air to push against.
    Y. travels forward as a result of the backward blast.
    Z. always travels too fast to be affected by gravity.

    (A) X.   *B.*
  ★(B) Y.
    (C) Z.
    (D) X and Y.
    (E) Y and Z.

SECTION 14. Transport Systems

157. P is the part of the tooth called the   *A.*

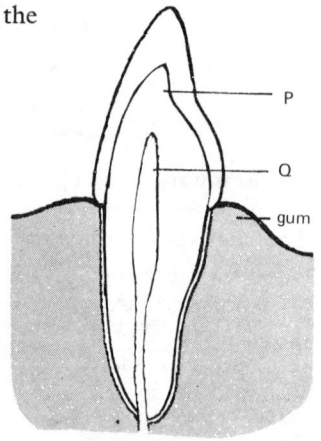

FIGURE T.25

    (A) enamel,
  ★(B) dentine,
    (C) root,
    (D) pulp cavity,
    (E) crown.

158. Q is the part of the tooth called the

    (A) enamel,   *A.*
    (B) dentine,
    (C) root,
  *(D) pulp cavity,
    (E) crown.

159. A dentist is fixing false teeth to a dental plate. If

X = incisors
Y = molars
Z = canines

in which order would he fit them (front, side, back)?

    (A) X Y Z.   *C.*
  *(B) X Z Y.
    (C) Z X Y.
    (D) Y Z X.
    (E) Z X Y.

160. Which of the following statements refers to incisors? They are

  *(A) used for cutting up food,   *A.*
    (B) used for grinding up food,
    (C) used for stabbing food,
    (D) sharply pointed in shape,
    (E) grooved on the surface.

161. Which of the following possess only incisors and molars?

  *(A) Rabbits.   *B.*
    (B) Earthworms.
    (C) Locusts.
    (D) Mussels.
    (E) Cats.

162. Which of the following is the best source of protein?

    (A) Bread.   *A.*
    (B) Cabbage.
    (C) Chocolate.
    (D) Apples.
  *(E) Fish.

163. Which of the following is the best source of starch?

*(A) Bread.   *A.*
 (B) Butter.
 (C) Jam.
 (D) Egg.
 (E) Bacon.

164. When a drop of milk was placed on a filter paper, an oily spot was
     formed. This shows that milk contains

 (A) starch,   *B.*
 (B) glucose,
 (C) protein,
*(D) fat,
 (E) water.

165. Fehling's solution turns red in glucose.
     A chemist spits (saliva) into a test tube containing starch solution.
     Drops are removed and tested as shown:

	Fehling's solution	Iodine
0 minutes	blue	blue
5 minutes	pink	pale blue
10 minutes	red	unchanged

The results show that saliva

*(A) changes starch into glucose,   *D.*
 (B) changes glucose into starch,
 (C) and Fehling's solution do not react,
 (D) and iodine do not react,
 (E) must contain glucose.

166. When glucose is burned in air, the gases produced are

 (A) nitrogen and water vapour,   *A.*
 (B) oxygen and carbon dioxide,
 (C) hydrogen and carbon dioxide,
 (D) oxygen and water vapour,
*(E) carbon dioxide and water vapour.

167. The importance of glucose to humans as a food is the formation of

 (A) oxygen,   *A.*

(B) carbon dioxide,
*(C) energy,
(D) water,
(E) vitamins.

168. The most important foods for growth and repair are

(A) fats,   *A.*
*(B) proteins,
(C) carbohydrates,
(D) vitamins,
(E) mineral salts.

169. Which of the following foodstuffs would a trainer of top class racehorses use as a balanced diet for his horses?

	% Protein	% Carbohydrate	% Roughage
(A)	80	20	0
(B)	20	80	0
*(C)	40	40	20
(D)	20	30	50
(E)	30	20	50

C.

170. During digestion food becomes soluble and passes from the digestive system into the blood. This mainly takes place in the

(A) mouth,   *A.*
(B) gullet,
*(C) small intestine,
(D) large intestine (colon),
(E) rectum.

171. Which of the following plays no part in our digestion?

(A) Salivary glands.   *A.*
(B) Pancreas.
(C) Gall bladder.
*(D) Appendix.
(E) Liver.

172. Which of the following is the correct order of parts in the digestive system, starting with the mouth

(A) gullet, small intestine, stomach, colon,   *A.*
*(B) gullet, stomach, small intestine, colon,
(C) stomach, gullet, colon, small intestine,

(D) stomach, small intestine, colon, gullet,

(E) small intestine, gullet, stomach, colon.

173. While we eat, some digestive fluid is passed into the mouth from the

(A) gall bladder,   A.

(B) pancreas,

(C) liver,

★(D) salivary glands,

(E) gullet.

174. Which of the following is an excretory organ?

★(A) Lungs.   A.

(B) Liver.

(C) Colon.

(D) Stomach.

(E) Heart.

175. Which of the following substances is removed from the blood as it passes through the kidney and stored in the bladder?

(A) Oxygen.   A.

★(B) Urea.

(C) Glucose.

(D) Carbon dioxide.

(E) Starch.

176. Which of the following statements is true? The skin

X: helps to regulate body temperature.

Y: is one route by which waste materials leave the body.

Z: produces the same amount of sweat all over the body.

(A) X only.   A.

(B) Y only.

★(C) X and Y only.

(D) X and Z only.

(E) X, Y and Z.

177. Which of the following is an example of ELIMINATION rather than of EXCRETION?

★(A) Leaf fall.   B.

(B) Formation of pigments, gums or resin in plants.

(C) Removal of urea from the blood by the kidneys.
(D) Storing of urine in the bladder.
(E) All of these.

178. Identical twins X and Y are found to have a pulse rate of 80.
Twin X, after running for five minutes, is found to have a pulse
rate of 100. Both twins then blow into beakers of lime water, and
twin Y takes far longer to turn the lime water milky. This experiment shows that increased exercise

(A) increases heart beat only,   *D.*
(B) decreases heart beat only,
★(C) increases heart beat and carbon dioxide output,
(D) decreases heart beat and carbon dioxide output,
(E) is not good for twin X.

179. Which of the following would NOT be found in human blood?
(A) Carbon dioxide.   *B.*
(B) Oxygen.
(C) Urea.
(D) Sugar.
★(E) Roughage.

180. The main function of haemoglobin in the red blood corpuscles
is to

(A) help in the formation of clots,   *A/B.*
(B) distribute heat,
★(C) carry oxygen round the body,
(D) destroy bacteria,
(E) carry glucose molecules round the body.

181. Digestion is the breaking down of large food molecules into smaller
ones. The main purpose of this is to

★(A) make the food soluble,   *A.*
(B) use up the digestive enzymes,
(C) make the food slide down the intestine,
(D) make the food insoluble,
(E) break down the roughage.

182. Which of the following is TRUE? Sweat
(A) is given off equally all over the body,   *B.*
(B) is a measure of water intake,

   (C) is a measure of body temperature,
⋆(D) consists of water and other waste products,
   (E) comes from veins in the skin.

183. Which of the following is NOT used for excretion or elimination of
waste products?

   (A) Skin.   *A.*
   (B) Lungs.
   (C) Kidneys.
   (D) Colon.
⋆(E) Liver.

184. The upper jaw of an animal is found with these teeth:

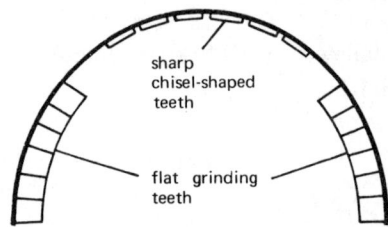

FIGURE T.26

meat eater = carnivore
grass eater = herbivore
meat + grass eater = omnivore

The animal could be

   (A) carnivore only,   *C.*
⋆(B) herbivore only,
   (C) omnivore only,
   (D) carnivore or omnivore,
   (E) herbivore or omnivore.

185. The skeleton of an unknown animal is dug up in a swamp. Which
of the following would tell us whether or not it hunted other
animals?

   (A) The length of its legs.   *C.*
   (B) The size of its body.
⋆(C) The shape of its teeth.
   (D) Whether or not it had horns.
   (E) Its number of teeth.

SECTION 15. Electricity and Magnetism

186. Which one of the following substances is regarded normally as a conductor of electricity?

   (A) Pure water.   *A.*
  ★(B) Iron.
   (C) Air.
   (D) Plastic.
   (E) Wood.

187. Current flows in the circuit shown.

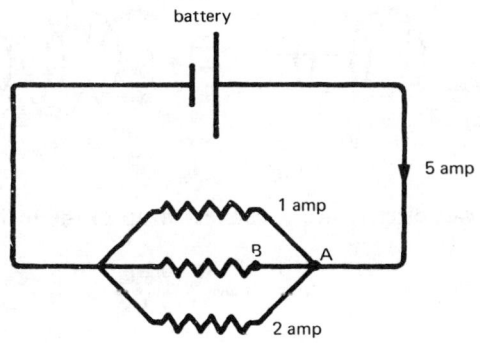

FIGURE T.27

The current flowing in wire AB is

   (A) 1 A.   *B.*
  ★(B) 2 A.
   (C) 3 A.
   (D) 5 A.
   (E) 8 A.

188. A 2 kW electric fire is used for 2 hours every evening for a week. How many units (kWh) of electricity are used?

   (A) 7.   *B.*
   (B) 14.
   (C) 21.
  ★(D) 21.
   (E) 35.

189. Which one of the following diagrams shows the correct arrangement of the lines of force produced by an electric current flowing through a coil of wire?

A.

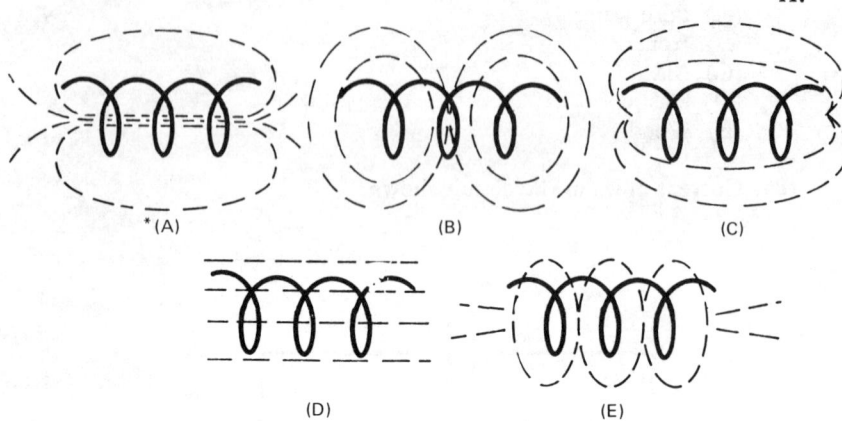

FIGURE T.28

190. A 1 kilowatt electric fire is connected up to the mains voltage of 250 volts. If

$$\text{current} = \frac{W \text{ (watts)}}{V \text{ (volts)}}$$

which of the following fuses would it be best to use?

(A) 1 amp only.   *B.*
(B) 2 amp only.
★(C) 5 amp only.
(D) 13 amp only.
(E) Any of these.

191. When the current is switched on, the shadow of a cross appears on the front of the glass tube at A.

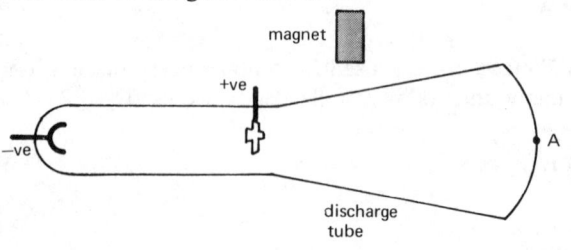

FIGURE T.29

When a magnet is brought near to A as shown, the shadow of the cross will

(A) get bigger,   *A.*
(B) get darker,
(C) get smaller,
(D) not be affected,
*(E) be deflected.

192. Six plotting compasses with their north poles pointing north are placed round a wire as shown.

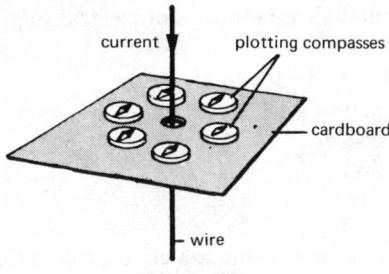

FIGURE T.30

When electric current is passed down the wire, the north poles of the compasses will

(A) keep pointing north,   *B.*
*(B) point clockwise round the wire,
(C) point south,
(D) point in towards the wire,
(E) point away from the wire.

193. The diagram shows a coil connected to an ammeter and a magnet inserted in the coil.

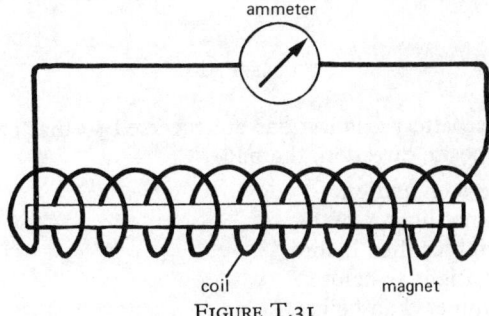

FIGURE T.31

Which one of the following statements is correct? The ammeter will show NO electric current when the magnet

  ★(A) and the coil both remain still,   *B.*
     (B) is moved slowly in and out of the coil,
     (C) is moved quickly in and out of the coil,
     (D) is kept still and the coil is moved backwards and forwards quickly,
     (E) is kept still and the coil is moved backwards and forwards slowly.

194. In which of the following pieces of electrical apparatus is a diode used?

     (A) Dynamo.   *A.*
     (B) Electric motor.
     (C) Electric cooker.
     (D) Electric fire.
  ★(E) Radio.

195. Direct current is passed round a circuit from a battery via a diode to light a bulb.

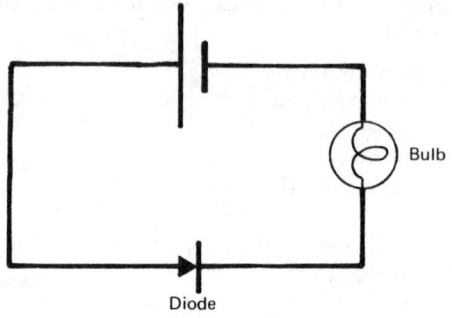

FIGURE T.32

When the battery connections are reversed so that current flows in the opposite direction, the bulb

     (A) flickers on and off,   *C.*
  ★(B) does not light,
     (C) is brighter than before,
     (D) is as bright as before,
     (E) is dimmer than before.

196. Which of the following statements about alternating currents are true?

    X: they change direction regularly.
    Y: they change strength regularly.
    Z: they always produce a large number of watts.

  (A) X only.   *A.*
⋆(B) X and Y.
  (C) X and Z.
  (D) Y and Z.
  (E) X, Y, and Z.

197. Wire XY passes mid-way between the poles of a strong magnet.

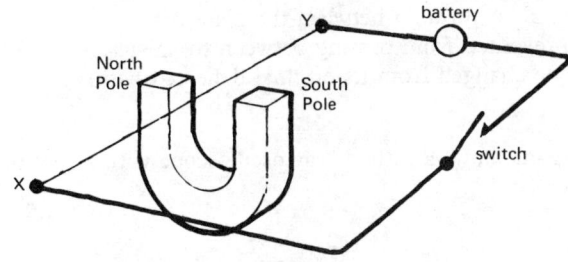

FIGURE T.33

When the switch is put on, wire XY will

  (A) be attracted to the north pole,   *B.*
  (B) be attracted to the south pole,
  (C) remain mid-way between the poles,
  (D) not carry current due to the effect of the magnet,
⋆(E) be pushed upwards away from the magnet.

198.

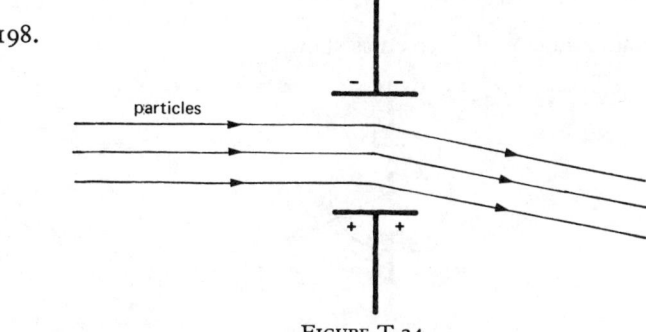

FIGURE T.34

The above diagram shows what happens when a beam of particles is passed between two electrically charged plates.

This shows that the particles could be

⋆(A) all negatively charged,
 (B) all positively charged,
 (C) a mixture of negatively charged and uncharged particles,
 (D) a mixture of positively charged and uncharged particles,
 (E) a mixture of positively, negatively and uncharged particles.

199. In the diagram for question 198, when the quantity of electrical
     charge on the plates is increased the particle beam will

 (A) be deflected less,   C.
⋆(B) be deflected more,
 (C) go straight through between the plates,
 (D) be prevented from passing between the plates,
 (E) not be changed from its original deflected path.

200. The diagram shows a cathode ray oscilloscope with a spot formed
     on the screen.

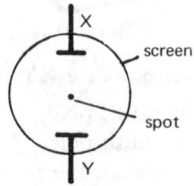

FIGURE T.35

To obtain a line on the screen as shown

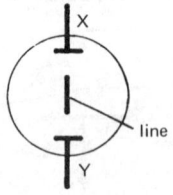

FIGURE T.36

the charged plates must be adjusted so that

(A) X = positively charged, Y = negatively charged,   *D.*
(B) X = negatively charged, Y = positively charged,
(C) X and Y are slowly changed back and forward from + to −,
★(D) X and Y are rapidly changed back and forward from + to −,
(E) X and Y are both negatively charged.

201. Which of the following statements about electricity in the home is
FALSE?

(A) Household appliances are connected in parallel.   *A.*
(B) Mains electricity is alternating current.
(C) Electricity is sold in units of kilowatt hours.
(D) Different households require different amounts of electrical
power.
★(E) Fuses convert alternating current into direct current.

202.

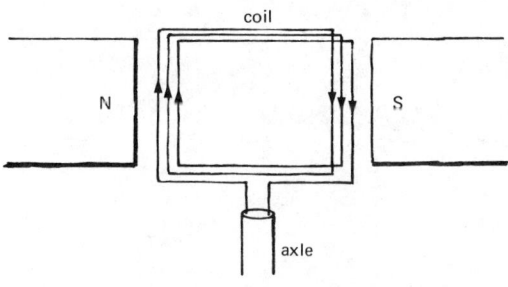

FIGURE T.37

The diagram shows an electric motor with a coil of wire carrying
electric current between the poles of a magnet. In which of the
following will the axle turn fastest?

(A) Strong magnet, weak current, many turns in the coil.   *B.*
(B) Weak magnet, weak current, many turns in the coil.
(C) Strong magnet, weak current, few turns in the coil.
(D) Weak magnet, strong current, few turns in the coil.
★(E) Strong magnet, strong current, many turns in the coil.

Appendix 2

**Cross references between experiments in the Pupils'
Textbook and the Science Worksheets**

Science Worksheets for the Scottish Integrated Science Course are
published by Heinemann Educational Books. The following list gives
the number of the Worksheet on which each experiment in the Pupils'
Textbook appears. There are, of course, more experiments in the book
than on the Worksheets.

References are given as follows:

Under the column headed 'Pupils' text' the experiments are numbered
as in the text. Thus 9.8 means Experiment 9.8 in Unit 9.

The corresponding worksheet references are given in the form 9/4.1
meaning the worksheet for section 9, sheet 4, frame 1.

Unit 9

Pupils' text	Worksheet	Pupils' text	Worksheet	Pupils' text	Worksheet
9.1	—	9.6	9/2	9.11	—
9.2	—	9.7	9/3	9.12	—
9.3	—	9.8	9/4.1	9.13	—
9.4	9/1.1	9.9	9/4.2	9.14	9/5
9.5	9/1.2	9.10	9/4.3	9.15	9/6

Unit 10

Pupils' text	Worksheet	Pupils' text	Worksheet	Pupils' text	Worksheet
10.1	10/1.1	10.9	10/3.5	10.18	10/4.5
10.2	10/1.3	10.10	10/3.6, 8	10.19	10/4.6
10.3	10/2	10.11	—	10.20	10/5.1
10.4	—	10.12	—	10.21	10/6.1, 2, 3
10.5	—	10.13	10/5.1, 2	10.22	10/6.4, 5
10.6	—	10.14	10/5.1	10.23	10/7.1, 2
10.7	—	10.15	10/5.3	10.24	10/7.3
10.8	10/3.1, 4	10.16	10/4.1	10.25	—
		10.17	10/4.3		

Unit 11

Pupils' text	Worksheet	Pupils' text	Worksheet	Pupils' text	Worksheet
11.1	—	11.11	—	11.21	—
11.2	11/1, 1–4	11.12	—	11.22	—
11.3	11/2, 1–4	11.13	—	11.23	—
11.4	—	11.14	—	11.24	—
11.5	—	11.15	—	11.25	—
11.6	—	11.16	—	11.26	—
11.7	—	11.17	—	11.27	—
11.8	—	11.18	—	11.28	—
11.9	—	11.19	—	11.29	—
11.10	—	11.20	—	11.30	—

Unit 12

Pupils' text	Worksheet	Pupils' text	Worksheet	Pupils' text	Worksheet
12.1	—	12.15	—	12.29	—
12.2	—	12.16	—	12.30	—
12.3	—	12.17	—	12.31	—
12.4	12/1; 12/2.1	12.18	—	12.32	12/8.3
12.5	12/1; 12/2.2	12.19	—	12.33	12/10
12.6	—	12.20	12/6.2	12.34	12/11
12.7	—	12.21	12/6.1	12.35	12/10
12.8	—	12.22	—	12.36	12/12
12.9	12/1	12.23	12/6.3	12.37	—
12.10	12/4.7	12.24	12/7.2	12.38	—
12.11	12/5.1	12.25	12/7.3	12.39	12/13.1
12.12	12/5.3	12.26	—	12.40	12/13.1
12.13	—	12.27	—	12.41	12/13.2
12.14	—	12.28	12/8.1	12.42	—

Unit 13

Pupils' text	Worksheet	Pupils' text	Worksheet	Pupils' text	Worksheet
13.1	—	13.12	13/4	13.22	—
13.2	—	13.13	13/5.1, 3	13.23	—
13.3	—	13.14	13/5.2	13.24	—
13.4	—	13.15	13/6	13.25	—
13.5	—	13.16	13/6	13.26	13/7
13.6	—	13.17	—	13.27	13/8
13.7	—	13.18	—	13.28	—
13.8	—	13.19	—	13.29	—
13.9	13/1.1	13.20	—	13.30	—
13.10	13/1.3; 13/2	13.21	—	13.31	—
13.11	13/3				

Unit 14

Pupils' text	Worksheet	Pupils' text	Worksheet	Pupils' text	Worksheet
14.1	14/2	14.11	—	14.22	14/10.2
14.2	14/3; 14/4	14.12	—	14.23	14/10
14.3	—	14.13	—	14.24	14/11;
14.4	—	14.14	14/6		14/12
14.5	—	14.15	—	14.25	14/13
14.6	—	14.16	—	14.26	14/13
14.7	—	14.17	14/8.1, 2, 3	14.27	—
14.8	—	14.18	—	14.28	—
14.9	14/5	14.19	14/9.1	14.29	14/14
14.10	—	14.20	—	14.30	14/14
		14.21	14/9.2	14.31	—

Unit 15

Pupils' text	Worksheet	Pupils' text	Worksheet	Pupils' text	Worksheet
15.1	15/1	15.9	—	15.17	15/10
15.2	15/2	15.10	—	15.18	—
15.3	15/3	15.11	—	15.19	15/11
15.4	15/4	15.12	15/6	15.20	15/12
15.5	—	15.13	15/7	15.21	—
15.6	—	15.14	15/8.1	15.22	—
15.7	—	15.15	15/8.2	15.23	—
15.8	—	15.16	15/9	15.24	—